WISE WORDS & COUNTRY WAYS

for House and Home

In loving memory of Donald, with
whom I shared both house and home.

WISE WORDS & COUNTRY WAYS
for House and Home

RUTH BINNEY

D&C
David and Charles

A DAVID & CHARLES BOOK
Copyright © David & Charles Limited 2009

David & Charles is an F+W Media Inc. company
4700 East Galbraith Road
Cincinnati, OH 45236

First published in the UK in 2009

Text copyright © Ruth Binney 2009

A catalogue record for this book is available from the British Library.

ISBN-13: 978-0-7153-3284-9 hardback
ISBN-10: 0-7153-3284-8 hardback

Printed in China by RR Donnelley
for David & Charles
Brunel House   Newton Abbot   Devon

Commissioning Editor: Neil Baber
Editorial Manager: Emily Pitcher
Editor: Verity Muir
Project Editor: Beverley Jollands
Designer: Victoria Marks
Production Controller: Alison Smith

Visit our website at www.davidandcharles.co.uk

David & Charles books are available from all good bookshops;
alternatively you can contact our Orderline on 0870 9908222 or write
to us at FREEPOST EX2 110, D&C Direct, Newton Abbot, TQ12 4ZZ
(no stamp required UK only); US customers call 800-289-0963 and
Canadian customers call 800-840-5220.

CONTENTS

INTRODUCTION

Home, it is said, is where the heart is. A truism indeed, and one with which
I concur totally. I love my home dearly but would be the first to confess that
housework – as opposed to home making – is not my favourite occupation,
unless it involves serious blitzing such as turning out an entire room for spring
cleaning. What is important to me is that my home is warm, comfortable and
always welcoming.

The home with all mod cons that I enjoy today is very different from those I grew
up in. These, which were largely unmodernized houses of a boys' boarding school,
were unbelievably hard to run. Even with the advantages of some domestic help, my
mother had to work inordinately hard to keep everything in order, yet she still had
the time and energy to entertain regularly. Even in 1965 when I had my own first
home, a flat in North London, there was no refrigerator; the spin dryer and twin-tub
washing machine were luxuries indeed.

The immediate post-war era of my childhood made frugality a necessity and
wastage anathema. This explains, at least in part, why it is a relief to see the world
turn full circle, so that it has now become totally acceptable to be careful with
things, to clean, renovate and repair rather than to throw things away, or at the very
least to send them to be recycled or donate them to a charity shop. In the kitchen,
leftovers are once again becoming bona fide ingredients. In my family the art of
the *réchauffé* is one that has been well nurtured over the years. My late husband
Donald's lamb bake, made from cold roast meat, was legendary.

In writing this book it has been enormous fun to look back and discover how
home makers of the past, particularly in the Victorian era but also in the 1920s,
30s and 40s, looked after their homes and possessions. Without access to modern
cleaning materials such as detergents, and with homes not only cluttered with
ornaments of every description but with every room warmed by open coal fires,
keeping everything free of dust and dirt was a huge task. Laundry took almost the
whole week and, without the benefit of refrigerators, fresh food needed to be bought
virtually every day. Entertaining in the home, whether for meals or overnight, was
usually formal and required a huge amount of preparation and effort. The only
people who ate supper around the kitchen table were the servants.

In exploring the homes of past times I have been fortunate to acquire, or have access to, some remarkably comprehensive works, including *Enquire Within Upon Everything* (mine is the revised 89th edition of 1894); four volumes of *Cassell's Household Encyclopedia*, subtitled *A Complete Encyclopedia of Domestic and Social Economy and forming a Guide to Every Department of Practical Life*, published between 1869 and 1871; *The Concise Household Encyclopedia* of the early 1930s; *Our Homes and How to Make Them Healthy*, a compendium of articles published in 1883; *Mrs Beeton's Book of Household Management* of 1868; and bound volumes of late 19th-century magazines including *Home Chat* and *The Girl's Own Paper*. It is vital to point out, however, that I have not tested the remedies, treatments and recipes quoted from these and other sources, and that readers using them do so at their own risk.

My thanks, as always, are due in large part to the team at David & Charles, especially my editor Neil Baber, to my daughter Laura and her husband Lewis, and to my many supportive friends and family. Once again, Beverley Jollands has proved an invaluable help with the editorial detail. This book has given me a hugely enjoyable opportunity of looking back to the old ways of doing things and comparing them with those of today. I hope that it brings my readers equal rewards.

Ruth Binney
West Stafford, Dorset, 2009

CHAPTER 1

EVERYTHING IN THE HOME

A home, so the old saying goes, should have a place for everything and everything in its place. But there is much more to home making than this. A home needs to be furnished and decorated so that it is comfortable to live in, is not damp or draughty and is soundly built, for, as it has long been acknowledged, 'no amount of elegance can compensate for poor foundations'. The ideal home is also light and airy, and positioned so as to receive maximum sunlight. Not only does this make the interior more cheerful it also, it is traditionally believed, reduces the amount of illness suffered by the residents. Equally, in these days of high energy prices, daylight remains the most economical means of illumination.

As to furniture, this should be chosen so as to complement the architectural style of the building and be of an appropriate size for both the home and its occupants. In choosing any furniture, good workmanship remains a prime consideration. Buying poorly made items, even at low prices, will prove to be false economy. Antique pieces are fine, but always need to be carefully checked for damage before you buy, in case they harbour active woodworm.

Carpets and curtains, when well chosen, add much to the comfort and attraction of a room, and home makers of today are fortunate to be able to take advantage of a vast range of decorating materials and accessories. Long gone are the days when items such as clocks and mirrors could be afforded only by the wealthy.

Sufficient cupboard room is desirable in all departments of the home

Including ample wardrobe space in the bedrooms. It is said that no home can have too many cupboards.

'Closets and cupboards, as fixtures in and about the various rooms of a house, ought to be regarded,' says *Our Homes* of 1883, 'as fairly good investment of capital, inasmuch as they invariably add to the attractiveness of a house by tending to obviate the necessity of acquiring sundry articles of furniture.' Indeed they do, and, as the guide points out, they can easily be 'fitted in' to handy recesses. The first built-in cupboards were made during the reign of Queen Anne for the display of china and pottery. They were either fitted into the corners of a room or made to sit on either side of a fireplace. The oldest free-standing cupboards date from the Elizabethan period. Originally known as court or livery cupboards, they were the equivalent of the modern sideboard.

For the bedroom, the wardrobe is an evolution of the 17th-century linen press, so named because it contained sliding trays designed to hold folded linen. In about 1800 hanging cupboards were added on either side and mirrors placed within. By the end of the century wardrobes were vast, often with a full-length mirror on the door.

A bacon cupboard is not really a cupboard at all but a type of settle with a backrest formed of a panelled cupboard and with drawers set below the seat. It was typically found in farmhouses from medieval times right up to the 19th century.

LINE A BATHROOM WITH TILES

A perfect way to protect the walls and floor from the effects of steam and damp, and a means of embellishing any bathroom or shower room. The decoration of walls with tiles is not new – ceramic tiles have been used in interior décor for at least 4,000 years.

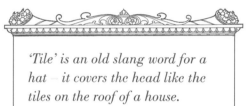

'Tile' is an old slang word for a hat – it covers the head like the tiles on the roof of a house.

Before the advent of the upstairs bathroom during the second half of the 19th century, the middle and upper classes washed in their bedrooms or adjacent dressing rooms. Water from the kitchen, very often cold rather than hot, was carried upstairs by servants and poured into basins or small tin baths. Bathrooms became possible only when plumbing systems could bring hot and cold running water to the upper floors of the house and were often created from converted dressing rooms.

Early bathrooms were decorated much as every other room; the floor might be covered with ceramic tiles, though for most people linoleum was a much more affordable option. By the beginning of the 20th century bathroom wallpapers were regularly protected with varnish and woodwork was treated with enamel paint, while the more well-to-do began to introduce wall tiles into the bathrooms of their newly built mansions.

The earliest recorded use of wall tiles was in Egypt around 2700 BC. Superb turquoise tiles lined chambers in the Step Pyramid at Saqquara, built for the Pharaoh Djoser.

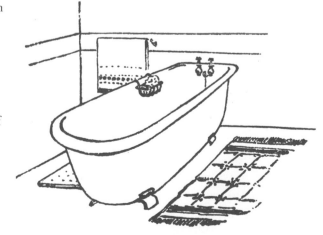

SHIELD A FIRE WITH A SCREEN

Or with a fire guard. Today, a screen is a good way of disguising a fireplace in summer, when the fire is not lit, but in winter a fire guard is a necessity for safety, especially if children are using the room.

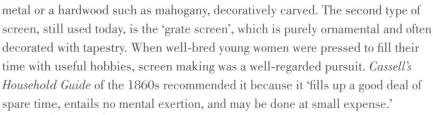

When open fires were the only means of heating the home, fire screens came in two types. The 'true' screen was a shield on a stand, which could be fixed in position to protect the face from a fire. Now rarely seen, it was usually made from metal or a hardwood such as mahogany, decoratively carved. The second type of screen, still used today, is the 'grate screen', which is purely ornamental and often decorated with tapestry. When well-bred young women were pressed to fill their time with useful hobbies, screen making was a well-regarded pursuit. *Cassell's Household Guide* of the 1860s recommended it because it 'fills up a good deal of spare time, entails no mental exertion, and may be done at small expense.'

In 18th-century Europe the fire screen desk was a popular item of furniture in prosperous homes, made in different designs for use by men and women. It was a miniature writing desk with a retractable fire screen at the back to protect the user's face from the heat, usually mounted on extendable metal rods forming an X shape. The desk was supported by an open trestle that exposed the feet to the warmth of the fire.

A firedog is a name for an andiron – one of a pair of horizontal iron bars, mounted on legs, that support the ends of logs in an open fireplace.

'A wire fire-guard, for each fire-place in the house,' says *Enquire Within*, 'costs little, and greatly diminishes the risk to life and property. Fix them before going to bed.' In Britain, the Children's Act of 1908 ruled that 'any person over the age of 16 years who has the custody, charge or care of any child over the age of 7 years is liable to a fine of £10 if he allows that child to be in any room containing an open fire grate that is not sufficiently guarded against the risk of being burnt or scalded …'

A DARK ROOM NEEDS A CHEERFUL WALLPAPER

Long acknowledged as an effective way of decorating a room, wallpapers are now available in a vast range of colours and designs. And happily for today's home makers, they no longer constitute a health hazard.

It is a myth that Napoleon died from chewing at arsenic-impregnated wallpaper during his imprisonment on the island of St Helena.

The earliest wallpapers – manufactured in the late 17th century as small block-printed sheets, rather like tiles – were used as substitutes for the thick, often dark wall coverings of textile or leather that had previously been used to decorate and insulate walls. In the 19th century, papers were printed with pigments made from substances such as lead and arsenic, and could be dangerous in damp rooms, when fungal growth on the wallpaper reacted with the dyes, leading to the emission of toxic gases. As late as 1895 *The Girl's Own Paper* warned: 'When having a house papered, make quite sure before they are put on the walls that they are not arsenicated … if you are doubtful about them submit a piece of each paper to a chemist or analyst and ask his opinion. Many cases of persistent illness have been traced to arsenic in the wall-paper, and it is not only present in green papers, but also in those of other colours.'

For adding light to a room, a pale paper is preferable to a dark one simply because it will reflect rather than absorb light. Pattern must also be considered, for as one 1930s guide explains, '… easily seen patterns may appear to reduce the size of a room.' To make a room appear larger, 'indefinite patterns in low tones tending towards greys will give the effect of space.' It also advises that 'bright yellow and orange can be employed for hall and staircase; amber, maize and golden tints for sitting room, lounge or dining room. Mottled effects resembling vellum and old parchment,' it adds, 'are very decorative.'

Before wallpaper paste was available for purchase, home decorators would make their own paste using a stiff batter made from plain flour and cold water, which was then mixed with ground alum and boiling water. It was strained through muslin and left to cool before use.

WALLPAPER HINTS AND TIPS

Old advice that still holds good today:

Warm tints such as yellow and red are helpful in north- and east-facing rooms, blue and grey for rooms with southerly aspects.

Always strip old paper off the walls before new is hung, including lining paper.

When buying paper, remember to allow extra for pattern repeats, and check that all the rolls are from the same printing batch.

Be sure to paste evenly and to every edge, and fold as you paste, pasted surface to pasted surface.

Begin papering at an angle of the room nearest the light.

Clean marks off wallpaper with a piece of stale bread or a soft, clean pencil eraser.

Make a wallpaper patch with torn edges for a perfect disguise.

Mirrors add light and space to a room

By the way in which they reflect light, mirrors do indeed give any room an airier, more spacious look. Once expensive luxuries and prized possessions, mirrors are especially well cared for by the superstitious.

Mirrors work best when they are hung at eye level and not crowded around with pictures and other ornaments; they are particularly useful in bringing extra illumination to dark hallways. A mirror will always look well over a fireplace (though its siting may be hazardous for the vain), a fact exploited by the Scottish brothers Robert and James Adam, whose elegant 18th-century designs for wall-to-ceiling fireplaces depended greatly on mirrors for the effects they created.

In public rooms the use of mirrors is chiefly decorative, whereas scrutiny of the person in a looking glass is the preserve of the bedroom or dressing room. Here, says *The House and Home Practical Book* of 1896, '[A] woman requires a long mirror. A man does not object to see the hem of his trousers. This is essential. This mirror may be a wardrobe panel, a cheval glass. If neither of these, on the bureau must depend the responsibility of furnishing a long mirror.' Also handy in the bedroom is an arrangement of three mirrors placed at angles so as to make it possible to see the back of the head – and therefore to arrange one's hair properly.

The original hand-held mirrors were probably pieces of polished bronze. In northern Italy, small hand-held mirrors were made around 5000 BC from the lustrous mineral obsidian, and by the Middle Ages polished silver mirrors were in use throughout Europe. Mirrors as we know them today, made of glass coated with mercury on the back to make it reflective, were developed by Venetian glassmakers on the island of Murano in the 16th century. The process they used remained a secret for over a century, keeping mirrors rare and expensive. Modern mirrors are coated with silver or aluminium instead of toxic mercury, but darkly sparkling antique mirrors give the most flattering reflections.

A mirror may be only as good as its frame. Those most prized by collectors have frames carved by craftsmen such as Grinling Gibbons (1648–1721) or are in gilt, in the French neoclassical style.

MIRROR, MIRROR

Good reasons to take care with mirrors:

Break a mirror and you will have seven years of bad luck.

A baby allowed to look into a mirror will have a troubled life.

It is bad luck to look into a mirror by candlelight.

If you look into a mirror for too long you are sure to see the Devil looking back at you.

During a thunderstorm, cover your mirrors with a cloth or turn them to face the wall, to stop them attracting lightning.

A GOOD MANTELSHELF IS IMPROVED BY A VELVET HANGING

… and a bad one is rendered endurable. This saying most certainly betrays its Victorian origins, since such decorations were beloved in an era when living rooms were bedecked with every kind of embellishment imaginable.

Cassell's Household Guide provides simple instructions for making such a hanging, for which the velvet is stretched over a shelf board placed over the mantelpiece itself. For fabric, it recommends Utrecht velvet and a decoration of 'a fine worsted fringe', adding that 'the mantle hanging always matches the window-curtains'.

The true mantelpiece is a part of the chimneypiece and may be made of seasoned wood or marble, cast iron or stone, according to the design of the fireplace. It may be left plain or, if wood, stained or varnished. In a well-appointed room the mantelpiece will be used to hold a clock and, possibly, candlesticks and other ornaments.

Though now a form of shelf, the mantelpiece originated as a simple hood to catch the smoke from an open fire and direct it up the chimney. With the evolution of interior design it became the focus of attention: it was enhanced with elaborate carvings and mouldings, and by the time of Louis XV it was often 'carried' by carved figures on either side of the fireplace.

In praise of the wooden mantelpiece, Robert W. Edis, a 19ᵗʰ-century writer on interior design, says that it is 'a pleasanter and more artistic object in a room than the usual boxed marble abomination'.

PICTURES, IF SMALL AND NUMEROUS, ARE BEST IN GROUPS

Enduringly good advice on the placement of pictures, which should always be sited where they can be best enjoyed. In older houses, picture rails remain desirable features for displaying works of art.

Small pictures dotted around a room create a spotted, untidy appearance. Equally, pictures should not be hung too high. As Mrs Caddy says in her *Household Organization*, '… few of us tower above six feet, and it is easier for those who do so to stoop, than for the rest of us to stand on tip-toe.'

Robert W. Edis, writing in *Our Homes*, advocates 'the division of the wall into two spaces, the one next the ceiling being made as a broad frieze distempered to a uniform shade and treated with good figure subjects in oil or stencil decoration, or covered with a paper of some "all over" pattern of a much lighter tone than the general wall-surface or lower portion of the room.' He suggests creating a picture rail using 'three quarter inch gas piping' if no such moulding already exists.

Choosing where to place pictures will depend on the positions of the windows and the way in which the room is lit, since good light enhances the looks of any form of artwork. Pictures are best kept away from fires and radiators and, especially if valuable, out of direct sunlight, which might damage them or cause them to fade.

The practice of covering pictures with glass was not always appreciated when introduced in the 19th century. Charles Dickens remarked how, having heard much about the Chandos portrait of Shakespeare on show at the National Portrait Gallery, he placed himself in front of it and saw nothing but an exquisite portrait of himself.

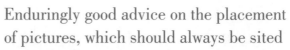

It is an old superstition that when a picture leaves the wall someone then receives a call.

No Household Article More Conspicuously Unites Useful And Ornamental Qualities Than The Clock

Clocks are, indeed, excellent additions to the home, whether modern battery-driven designs or handsome grandfather clocks and other heirloom timepieces that need to be regularly wound.

As well as being good looking, a clock needs to tell the time accurately for, as *Cassell's Household Guide* points out, 'were it deficient in correctness' it would 'instead of being the monitor of the household' be 'a false guide to all the arrangements on which the regularity of the family depends.'

The earliest mechanical clocks – typically those placed in church towers – were driven by weights turning a wheel controlled by an escapement mechanism. Small clocks, designed to be moved from place to place, were driven either by weights or by coiled springs. From the 17th century spring-driven clocks incorporated a balance spring to compensate for the decreasing force of the mainspring as it uncoiled, and a short 'bob' pendulum was added to increase accuracy. When small spring clocks first appeared in English houses in the early 17th century they were known as birdcage or lantern clocks.

Longcase clocks became possible after Galileo had come to understand the principle of the pendulum in the 1580s, and they were as much pieces of domestic furniture as timepieces. Most valuable of all are those made in London in the 'classic period' between 1660 and 1720. The longcase clock is often known as the grandfather clock; it gets this name from the popular song 'My Grandfather's Clock' written in 1876 by the American Henry Clay Work, which includes the lines:

My grandfather's clock
Was too large for the shelf,
So it stood ninety years on the floor;
It was taller by half
Than the old man himself,
Though it weighed not a pennyweight more.
It was bought on the morn
Of the day that he was born,
And was always his treasure and pride;
But it stopped short
Never to go again,
When the old man died.

REGARD WITH RESPECT

The value of the clock is reflected in many old sayings and superstitions:

Never stand a clock so that it is facing the fire, or the fire will go out.

It is said that a clock will stop at the moment of a death.

If you talk while a clock is striking you will have bad luck.

Although we can learn from experience it is impossible to put back the clock.

THE EXCELLENCE OF A CARPET DEPENDS ON THE THICKNESS OF THE PILE

Pile is certainly the test of a carpet, whether it is made by hand or machine. Other important considerations when choosing a carpet are the quality of the yarn, the colour and the pattern.

The key to assessing the pile of a handmade carpet is the number of knots per square centimetre. In a good Persian carpet, made from fine silk yarn, the figure will be 50 at the minimum, whereas in a quality modern machine-made tufted carpet such as an Axminster or Wilton, made from thicker yarn, there will be about 12 knots per square centimetre.

For a handmade carpet, the warp threads are set up on a loom into which are woven rows of looped knots, which are then cut to make the pile. The knots are secured by rows of weft threads. They may be symmetrical (Turkish style) or asymmetrical (Persian style).

Carpet making was probably begun by nomadic shepherds of Central Asia in the second millennium BC. Their creations would have been used as blankets and for covering walls and doorways. The oldest surviving knotted carpet, dating from the fifth century BC, is the Pazyryk Carpet, which was excavated from a Scythian royal tomb in Siberia in 1949. Though faded, its many colours and intricate pattern, including borders depicting deer and horsemen, can still be discerned.

The skill of carpet weaving was brought to Britain in the early 16th century, probably by Flemish Calvinists fleeing persecution. Norwich was one of the

earliest centres of carpet making, as was Wilton
in Wiltshire. In 1755, Thomas Whitty opened a
wool carpet manufacturing company in the Devon
town of Axminster, which quickly became very
successful, but the true centre of carpet making was
Kidderminster in Worcestershire, where the first
machine-made carpets were manufactured in 1735.

For choosing a patterned carpet, *Enquire Within*
gives advice that still holds good: 'For a Carpet to be
really Beautiful and in good taste, there should be, as
in a picture, a judicious disposal of light and shadow,
with a gradation of very bright and of very dark tints,
some almost white, and others almost or quite black.'

CURTAINS ADD MUCH TO THE COMFORT AND ADORNMENT OF A ROOM

Indeed they do, for curtains provide insulation as well as adding
to the privacy of the occupants. But because of the propensity of
curtains to gather dust, doctors and health gurus once favoured
blinds over curtains, especially in sickrooms.

The first curtains probably did not hang at windows. Rather, in the medieval age,
when several members of a family would share a sleeping space, they were placed
around a bed for warmth and privacy. Heavy fabric might be stretched over wooden
frames, or fenestrals, to cover the windows, but shutters mounted on the insides
of the windows were more usual. An inventory of 1509, for the London house of
Edmund Dudley, contains the earliest written reference to window curtains.

These would have hung from a rod secured above the window, but even by 1700 such curtains would still have been considered a luxury. By the 19th century, curtains in every kind of fabric, from chintz to cretonne, were being used. In addition, windows were hung with muslin curtains, which, although they restricted the amount of light entering the room, were deemed essential to keep out the prying eyes of the neighbours. Even thick curtain fabrics would be lined, to add to their warmth-giving qualities, and curtains would be made long enough to drape over the floor to help exclude draughts.

Reflecting on the appearance of curtains, Jane Ellen Panton in *From Kitchen to Garret* is strident – even snobbish – in linking the choice of fabrics to furnish a home with the character of its occupants: 'The carefully and prettily and tidily arranged curtains,' she says, 'tell me at once of the pleasant folk I shall find inside; just as surely as the dirty, untidy muslin or the gorgeously patterned, expensive, and pretentious curtains warn me against the slattern, or the vulgarian with whom I have nothing in common, should I ever have the bad fortune to have to enter behind those warning marks …' For her, 'soft Madras or delicate lace' indicate 'an artistic mistress with whom I shall, I know, spend many pleasant hours.'

A 'curtain-lecture', says Samuel Johnson in his 1755 *Dictionary*, is a reproof given by a wife to her husband in bed.

A BIG SETTEE IS A MOST USEFUL POSSESSION

And an ideal piece of furniture for a drawing room. Today the words 'settee', 'sofa' and 'couch' are used interchangeably, but they originate, in fact, from different types of seating.

The settee is named from its forerunner the settle, which was a wooden bench, usually with arms and a high back (essential to exclude the draughts common in medieval halls), and with a seat long enough to accommodate three or four sitters. It would often have winged ends and sometimes a wooden canopy, also for increased comfort. The oldest settles, dating from the 16th century, have more in common with church pews than with home comfort. As upholstered furniture was developed so the settle had cushions added.

A design that retained its popularity well into the 1920s was that of the Queen Anne period, with plain cabriole legs, arms shaped like shepherds' crooks and with the padded back and seat upholstered in tapestry.

In the distant past the couch was an item of furniture found in the ancient Roman dining room, or *triclinium*. Three couches would be arranged around three

sides of a low table and on them men reclined while eating; until the late Empire period women ate separately, and would sit on upright chairs.

The term 'sofa' comes from the Arabic *suffah*, meaning cushion, and originally referred to the raised section of the floor, furnished with rugs and cushions, which was set apart for a council. The first modern sofas were those by craftsmen and designers such as Thomas Chippendale and Robert Adam in the late 18th century.

Day beds first came into use in the late 17th century, and the earliest designs were essentially elongated chairs. By the Victorian era they had become sofas with drop ends, allowing them to be used for either sitting or reclining. They often had seats divided into three fully adjustable cushions, with a movable headrest and arm supports.

COMFORT IN MANY DESIGNS

Some more variants on the settee include:

CONFIDANTE OR CONVERSATION SEAT – a sofa with an S-shaped top rail, allowing two people to sit beside each other while facing in opposite directions.

CHESTERFIELD – a Victorian well-padded sofa, often upholstered in leather and usually buttoned, with back and arms of the same height.

SOCIABLE – a variant of the confidante with swivelling seats.

CHAISE LONGUE – a cross between a sofa and a chair with a whole or partial back and a long seat for reclining.

OTTOMAN, BORNE OR TURKEY SOFA – a low upholstered seat without arms, which may or may not have a back.

THE TEST OF A GOOD CHAIR LIES IN ITS WORKMANSHIP

Since a chair needs to be both strong and comfortable, good workmanship is indeed important, whether you are choosing dining chairs or armchairs. In the home, or when visiting, the superstitious will always treat chairs with respect.

The swivel chair, invaluable at the study desk as well as in the office, was the invention of American President Thomas Jefferson.

The seat is arguably the oldest piece of furniture, and the form of the chair has changed little since the ancient Egyptians made chairs with four straight legs and a high, upright back. Styling evolved with the ancient Greeks, who favoured *klismos* chairs with curved legs and concave backs. In the Regency period the Greek style was revived by neoclassical designers, becoming the height of fashion; it came into vogue again in the Art Deco era, and chairs of this shape are still made. Wicker chairs, not dissimilar to those produced today, originated with the Romans.

The first chairs were intended as seats for the high born and those in authority, literally raising them above lowlier mortals, who would sit on benches, or on simple stools, or even on the floor. Only with the Renaissance did chairs become suitable items of furniture for anyone who could afford them. Until the mid-17th century, when it became customary to upholster chairs with leather, velvet or silk, they were made of plain wood, usually oak.

The 18th century was the golden age of the chair, culminating in the Louis XVI chair with an oval back, ample seat, descending arms and round legs; such a chair would be upholstered in tapestry depicting scenes reminiscent of the paintings of the rococo artists of the era.

When testing a chair for comfort, one might bear in mind this paragraph from *Cassell's Household Guide* regarding easy chairs: 'There is more care required in the selection of these necessary articles than in almost any other. Some appear as if they were arranged rather for penitential chairs than anything else. The back aches, or the neck becomes stiff while sitting in them, although perhaps they look truly comfortable until they are tried.'

SIT WITH CARE

Some superstitions involving chairs:

It is unlucky to sit next to an empty chair.

If you overturn your chair as you get up you will never marry – or will fall ill.

A chair being passed over a table foretells an argument.

If you are out visiting, do not push your chair under the table when you rise at the end of a meal or you will never be invited to that home again.

Before sitting down to play cards, turn your chair around three times to ensure that you have good luck.

SELECT A TABLE ACCORDING TO ITS USE

… and the room in which it will stand. Different styles of tables – each a symbol of sharing and hospitality – are necessary for kitchen, dining room and drawing room, while special designs are handy for activities such as playing cards.

The oldest form of dining table was a plank resting on trestles. After each meal this assembly would be dismantled to allow space in the hall for dancing and other amusements. The monasteries were probably the first dwellings to have permanent dining tables – so-called refectory tables – but by the 17th century they had become commonplace in ordinary homes. In the same era tables began to be made with draw leaves, which allowed them to be expanded when necessary to accommodate extra diners.

The kitchen table is much more utilitarian, needing to double up as a chopping board. In large houses with servants it also served as their dining table. Many kitchen tables were made from pine, though some were in elm or apple wood. Whatever the wood it needed to be able to stand being scrubbed on a daily

basis. The first such basic tables date from the early 18th century.

Nests of occasional tables first appeared around the end of the 18th century. In the Victorian era the tops of these small decorative tables were often covered with mother of pearl or embellished with romantic scenes painted in oils. Many such sets were made of papier-mâché rather than wood. Equally useful in the drawing room is the small drop-flap table, or Pembroke table, supposedly devised in the 18th century by the Countess of Pembroke. A variation on this is the sofa table, which appeared in the 1790s. It could be placed in front of a seated person to be used for reading and writing.

Small folding tables, which could be erected when needed for playing cards or serving tea, became popular in the 18th and 19th centuries. The top of a card table is usually covered with baize and there are often wells at the sides for counters or chips. Such tables may also have drawers for storing cards and game pieces.

TABLE SAYINGS

At a round table there is no dispute of place (that is, no hierarchy of position).

A poor man's table is soon spread.

Lay an extra place at table for an absent friend.

The girl who sits on a table will never be married.

Wine makes all sorts of creatures at table.

The beggar sets his feet under another man's table.

A lantern on the table is death in the stable.

The dogs eat the crumbs that fall from their master's table.

OF ALL TYPES OF INTERIOR BLINDS, VENETIAN ARE THE BEST

Such blinds are certainly attractive, but not
the only good choice for a room. Roller blinds –
also called hollands – work well, as do Roman blinds,
which fall into decorative pleats as they are raised.

To keep the sun from a room and to diffuse the light, blinds are a much better
solution than curtains. Among the oldest of all blinds were those made from strips
of reed bound together, used by the Egyptians to shade their homes. The modern
equivalent is the pinoleum blind.

Venetian blinds, originally comprised of thin slats of wood (but now also
available in metal and plastic), were advertised as a novelty in London by John
Webster in 1767, but such blinds had actually been used for many centuries
before this in Japan. They are said to get their English name from the fact that it
was Venetian traders who introduced them to
Europe from Persia. In France they are still
known as *les Persiennes*.

*The holland blind is named from
a type of stiff woven linen from
which it was traditionally made.*

As to choice of colour, a blind needs to
fit with the scheme of the room, but also to
filter the daylight pleasingly. Mrs Caddy
had strong views about the subject. 'Red
tammy,' she said, 'enriches the colour of
the room, but it is not agreeable to sit long in a room filled with the flame-coloured
light, though this softens as the blinds fade, which they soon do. Yellow blinds are
very disagreeable, and tryingly sunny in summer. Blue are as unpleasantly cold,
and make people look like ghosts. White holland gives as soft a light as any, and if
carefully used the blinds will not go awry.'

ALL DRAUGHTS SHOULD BE WELL EXCLUDED

Ideally through the good construction of a home, both outside and in. Modern advances, such as the fitting of double glazing, have helped greatly in reducing draughts, making homes more comfortable and less expensive to heat.

Hermetically sealed double glazing, made from toughened, laminated glass, came into use in the 1960s.

As wind blows against the walls and windows of a house it seeks out any cracks and crevices, so causing draughts. And the more quickly it enters the worse a draught will be. Inside the house, draughts blow around ill-fitting doors and from beneath floorboards that are not snugly aligned and not covered with carpet. Windows are also a common source of draughts because as warm air comes into contact with a cold window it cools and sinks downwards. As it is replaced by more warm air a draught is created. Double glazing helps greatly in reducing this problem.

An old-fashioned open fire without its own vents will exacerbate draughts by pulling in air from across the room, especially from beneath a door. A space under a door can be filled with a draught excluder made like a sausage-shaped cushion. An amusing 'sausage dog' draught excluder can be easily made from one leg of an old pair of thick tights. Stuff it with old pieces of fabric, sew up the ends, then add felt ears, sew on buttons for eyes and embroider a mouth. You could even add a piece of red felt for a tongue.

Cassell's Household Guide offers mats as a solution. These, it says, 'ought to fit the doors exactly and completely. If they do not they are ornamental and not useful … the narrow mat, twelve inches wide, not only serves all purposes of use, but looks best in a limited space. Those of sheepskin are handsome and efficient; but for upper bedroom doors excellent mats may be made of cloth cuttings, sewn on to canvas in innumerable loops.' An alternative is to fit a purpose-made moulded strip to the base of the door.

TIPS AND HINTS FOR EXCLUDING DRAUGHTS

Fit draught-excluding strips on doors and windows.

Put covers over keyholes – and the letterbox.

Put good quality underlay beneath all carpets.

Block off any unused flues – but insert a small ventilator to prevent the accumulation of damp.

Never remove or obstruct air bricks in external walls, which admit air at low velocity: they are essential to keep air flowing, particularly under floors, and prevent problems such as dry rot.

Tighten up loose door latches.

Make sure curtains fit windows snugly and, if possible, have windows double glazed.

A SCRAPER AT THE DOOR KEEPS DIRT FROM THE FLOOR

A boot scraper is indeed a good way of saving mud and dirt from being brought into the house. Both outside and inside the door, a doormat is also useful for cleaning the shoes.

The scraper is definitely an item for the back, rather than the front, door. Made from metal, or wood with a metal scraping iron, if it is free standing it needs to be heavy enough to stay in position while it is being used. Georgian designs, with pleasing curves, are desirable items today. But a boot scraper will remove only the worst of mud and dirt. 'A good mud brush for boots,' says Mrs E.W. Kirk in her 1924 book of recipes and household hints, 'is made by glueing a strip of Brussels carpet to a convenient-sized piece of wood. No injury to leather, and less dust and labour than a brush.'

To be a doormat is to let other people exploit or walk over you. The Chat-Mat is a talking doormat that plays recorded messages when stepped on. It was the 1995 brainchild of Ralph Baer, the inventor of the first commercial video game system.

The archetypal doormat is made from coir, the fibres found between the husk and outer shell of a coconut. Before coconuts – and all their by-products – were brought to Europe in the 16th century, doormats would have been made from woven rushes or straw. The advantage of coir is that it is excellent at trapping dust and dirt. For cleaning, doormats can be shaken vigorously, beaten or vacuumed. You may want to encourage family members to remove their shoes (and certainly Wellington boots) before coming into the house, but no polite hostess would insist that her guests do so.

The doormat is also known as the welcome mat – and may even be emblazoned with the word 'Welcome' or some other greeting.

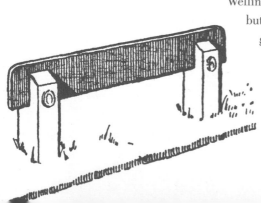

A BEDROOM WINDOW SHOULD OPEN AT THE TOP

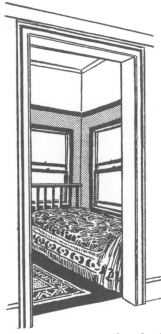

A saying that dates to a time when sash windows were the norm, the idea being to provide ventilation without the room being draughty. Whether the window should be left open at night remains a matter of debate.

Sash windows, introduced to England in the late 17th century, became popular from the early 1700s and by the Victorian era were virtually standard. The best type was the double-hung sash, which – as *Our Homes* of 1883, edited by Shirley Forster Murphy, explains – 'consists of two frames sliding vertically in parallel grooves. The upper sash is hung in the outer groove, to avoid making a lodgement for rain, which would be the case if the lower sash were outside. Each sash is hung by cords, which are taken over the wheels of pulleys, fixed as high up as the frame will allow, to which are attached iron or lead weights, which should be slightly lighter than the sashes.'

Regarding health, some say that the bedroom window should be kept open a little all day and also at night unless the weather is very cold. *Enquire Within*, however, is adamant that it 'should never be opened at night but may be left open the whole day in clear weather.' It also recommends, as substitute for a pane of glass in one of the upper lights of a chamber window, 'a sheet of finely perforated zinc', this being 'the cheapest and best form of ventilator.'

It was advised in Victorian times that if the window of a sickroom could be opened only at the bottom, a light shawl should be thrown over an invalid's face while the window was opened to refresh them.

EVERY HOUSEHOLD SHOULD HAVE A DRAWER FOR PAPER AND STRING

A saying that well reflects an age of thrift, when such resources were prized and when parcels had to be wrapped and tied without the aid of sticky-backed tape. In a truly frugal household crumpled paper would be smoothed with an iron before being reused.

String is essential for cat's cradle, a game in which two children wind it into patterns on their fingers, then pass it, in specific moves, back and forth between them. Probably one of the oldest of all games, different versions are played in cultures all over the world.

Instructing young ladies with 'household hints', *The Girl's Own Paper* of 2 November 1895 goes further. 'In every house,' it says, 'there should be a drawer in which string, scissors, nails, hammer and other small tools should be kept for immediate use. In another brown paper, neatly folded, should always be kept. Much trouble may be avoided,' it adds, 'if these things can always be found in the same place ready for use.' *Enquire Within* advises differently, suggesting that 'wrapping paper may be piled on the floor under a large shelf. It can be bought,' says the guide, 'at a low price by the ream at large paper warehouses.'

As well as parcelling, brown paper was used for lining drawers and for layering silks and ribbons to preserve their colours. And when, in the nursery rhyme, Jack and Jill fell down the hill, Jack 'went home to mend his head with vinegar and brown paper', reflecting the old use of paper soaked in vinegar as a poultice.

CHAPTER 2

CARE AND REPAIR

If, as the proverb goes, the want of thrift is the cause of misery, then it certainly pays dividends to take good care of all your possessions, including those that furnish and decorate the home. And caring means much more than cleaning. It means keeping and storing everything properly, keeping the home smelling sweet and free of cooking smells, and attending to annoyances such as smelling drains, dripping taps, curling rugs and squeaking doors.

A well-cared-for home is also free of pests. It has no cockroaches in its kitchen and is free from trouble with rats and mice. Being clean and well ventilated, it is most unlikely to harbour fleas. Equally, preventive measures are taken to deter moths from eating clothes and carpets. Any pests that do make their presence felt are dealt with swiftly and efficiently.

Rather than throwing things away if they show signs of wear, good home makers will take pains to see to renovations and repairs, or get an expert to do them. In this regard, many of the old ways are still the best. It still pays to keep shoes in shape with shoe trees, to shake feather pillows regularly and to revive oil paintings by cleaning off surface grime. In many respects, modern materials make it much easier to care for the home and its contents than in the past, but it is still worth knowing how to keep clothes well brushed and pressed, how to store silks that are worn only occasionally and how to cure annoyances such as squeaking doors and creaking stairs. In the kitchen, knives will, of course, be kept sharp, kettles free of limescale and larders well aired.

RUB A SQUEAKING DOOR WITH SOAP

Doors that squeak or do not shut properly are a constant
annoyance, as are floorboards that creak when stepped on.
For both complaints, simple remedies are usually the best.

The ideal way to apply soft soap to squeaky or creaky door hinges is on the tip of a
sharpened pencil. This has the added advantage that the graphite from the pencil
will provide extra lubrication. An old-fashioned product that also works well, if
you can get hold of it, is graphite grease, or you may want to try one of the modern
petroleum-based sprays. If none of these is successful, then the only remedy may be
to replace the door hinges altogether.

As for creaking floorboards – the bane of illicit lovers – the first thing to try is
moving the furniture so as to alter the pressure points on the floor. If this doesn't
work then you will need to check more
closely. If the floorboard is nailed down, it
is worth removing the nails and replacing
them with screws, which will hold the boards
in place more firmly. But be careful. The
screws you use must not be longer than
the nails you have pulled out, or you could
damage a water pipe or wire beneath – with
disastrous consequences. The ultimate
remedy may be the replacement of boards
that are beyond repair. Or it may be that the boards are laid too tightly together, in
which case dusting chalk or talcum powder between them may solve the problem.

*The proverb 'A creaking door
hangs long' is an expression
traditionally applied to invalids
who live on into old age despite
their ailments.*

Soft soap or soap jelly can easily be made at home from old scraps, as would
have been customary in wartime and in days when good soap was an expensive
luxury. Put a cupful of soap scraps in a pan with two cupfuls of boiling water and
heat until melted. Add a teaspoon of powdered borax, mixed to a paste with cold
water, stir well then leave the mixture to dry and set.

A CUP OF VINEGAR WILL STOP COOKING ODOURS FROM PENETRATING THE HOUSE

The original form of air purifier, though hardly a delightful smell. It was once common practice to set bowls of water around the house to absorb odours, while the pomander and pot pourri are old ways to disguise rather than dispel household smells.

When all rooms, including bedrooms, were heated with coal fires, cigars and cigarettes were smoked indoors and furnishings were rarely, if ever, cleaned, the home was a smelly place. Add to this effluvia from drains and smells from the kitchen, and it is not surprising that even leaving windows open overnight (an undesirable practice in winter when it was hard enough to keep warm) would not restore a room to total freshness.

One way of perfuming a room was with pot pourri made from mixtures of dried, scented flowers, herbs and leaves preserved with spices. A typical Victorian recipe reads: 'Take a handful each of lavender flowers, rose petals, sweet briar leaves, jasmine, rosemary and half a handful each of sweet geranium leaves, mint, thyme and lemon verbena. Put layers of these in a wide-mouthed glass jar with bay or rock salt between each layer. Cover for a month and after turning out the mixure add 1oz powdered orris root, a few drops each of oil of neroli, oil of musk and oil of cinnamon.'

Personal hygiene was once much less thorough than it is today. Even until the 1950s most adults would bath only once or at most twice a week, while clothes (and not just outer

It was once thought that a bowl of water placed under the bed would not only keep the room free of smells but would also deter bedbugs. Equally, cold water placed near the head of a restless sleeper was said to bring 'quiet and relief'.

garments) were worn for weeks on end before being laundered. To counteract body odours, the pomander originated as a means of personal perfumery. It gets its name from the French *pomme d'ambre*, meaning 'apple of amber' and was originally a perforated spherical container filled with ambergris or musk and hung on a chain around the neck or worn attached to the girdle. A modern version is made by sticking an orange all over with cloves then rolling it in a mixture of ground cimmanon, nutmeg and cloves. It is left to dry for at least a month, being turned in its spices daily, and can then be hung up with a pretty ribbon.

Any larder should be well aired

And also cool and even in temperature, a north-facing aspect being ideal. The original larder was not a general kitchen store cupboard but specifically a place for storing fresh food, in contrast with the pantry, which was for china, glass and silver, and the scullery, which was for messy food preparation and the scouring of pots and pans.

To keep it cool, the ideal larder would have had a good flow of air passing through it and would, in the Victorian home, be placed as far as possible from the heat of the kitchen range. As the guide *Our Homes* explains, '[It] may be a mere cupboard in size, but must be well ventilated into the outer air, and the ventilating must not be in close proximity either to the water closet or the dust-bin ...'

Shelves would be of marble or slate, and hooks from the ceiling held joints of meat. Instead of glass, the windows were usually covered with fine wire gauze to keep out flying insects; removable glass panes were often placed behind the gauze in winter.

Great care had to be taken to keep fresh food in good condition. Milk was customarily boiled before it had a chance to turn sour, sometimes with a little saltpetre dissolved in it. Vegetables were kept on the dry stone floor, where they were protected from frost. The meticulous cook, maid or housewife would wipe meat hooks each time they were used.

KEEPING THE LARDER CLEAN

Our Homes declares larder cleanliness as 'frequently neglected' and recommends the following routines:

DAILY:
Tidying of shelves.

Removal of all stale food not in good condition because 'decay in one thing will be communicated to others'.

Paying special attention to the isolation of 'mouldy cheese, rancid butter, sour milk and tainted meat'.

Separation of strongly flavoured ingredients such as bloaters and onions from butter or bread.

Prompt removal of dirty plates and dishes, and jugs and bowls that have contained milk or cream.

ONE OR TWICE A WEEK:
Scrub the shelves with sand and warm water; scour the floor.

THE CHEAPEST WAY TO MAKE A FLOOR LOOK ATTRACTIVE IS TO STAIN IT

A good remedy for a wooden floor, and one that is now both fashionable and attractive. What is most important is that the floor is thoroughly clean and level, and that any cracks between the boards are filled.

The Household Handy Book of 1933, recommending floor staining, says: 'For a shilling the boards of a fair-sized room can be done … The lino needed for the same purpose would cost anything from twenty shillings to three guineas, and a carpet would come to a great deal more.'

Before staining a floor it is essential to take a hammer and nail punch and sink in all the nail heads. Wide cracks and knot holes need to be filled – this can now be done with a proprietary paste, but the old way was to use papier mâché or putty. Any grease marks, which will stop a stain 'taking' well, need to be scrubbed with detergent. The floor can also be sanded – an operation that is vital if you need to remove any previous stain.

> When staining a floor remember to work towards the door so that you do not have to walk on treated boards.

Wood stain can be water or oil based, or combined with a varnish to give a shiny finish. When using the latter, it is well to remember that colours darken as additional coats are applied. Still in frugal vein, *The Household Handy Book* supplies several money-saving formulae, including, for a mahogany effect, bichromate of potash crystals dissolved in water; to imitate walnut, Brunswick black mixed with turpentine and a small quantity of varnish, and to imitate oak 'apply ordinary creosote', adding, 'This is very cheap, and is a splendid preservative for the wood. Some people object to the smell, however.'

ALWAYS CURE A DRIPPING TAP

Not just because it is an annoyance but to save precious water. The tap or faucet we now take for granted was hailed by the Victorians as a great household invention. And so it was, doing away with the necessity of drawing water from a well or from a communal water supply in the street.

A method of torture in which water is dripped slowly and continuously on to a person's forehead, allegedly driving them insane, was probably first described by the lawyer and doctor Hippolytus de Marsiliis in Italy in the 15th century. It is erroneously known as Chinese water torture.

Essentially, all taps work in the same way. A rotating handle opens and closes a valve inside the tap, allowing water to flow or cutting it off. A traditional spindle tap, operated by turning its head, is most likely to drip because it contains a worn washer. A less likely cause is damage to the valve seating. After turning off the water supply at the stopcock it is necessary to unscrew the tap cover and to remove the large headgear nut in order to locate the washer that lies within. If the valve that sits below the headgear nut is faulty, then it will be necessary to remove and replace the whole tap interior.

Modern taps operate using slotted ceramic discs, which rotate against each other to open or close the slots, through which the water flows. These discs are far more hard-wearing than rubber washers, but if they become scratched or worn the tap will start to drip. Replacement parts are, however, easy to source and fit.

A CREAKING STAIRCASE IS A GREAT ANNOYANCE

Indeed it is, and it is most likely to be a problem in an old home with a wooden staircase. Stair carpets also demand attention regarding their care and repair.

The noise of creaking stairs may be tolerated during daylight hours but becomes a particular nuisance at night. 'A good way to overcome the trouble,' says *The Housewife's Handy Book* of 1933, is to 'drive 1-inch nails through the upper part of each riser so that they slant to the tread … If about four nails penetrate each riser in this way, the treads will be held firmly and have no opportunity of creaking. Another better plan,' it adds, 'which is only possible when the under part of the stairs can be reached, is to cut a number of right-angled shapes out of a long plank of wood and to fit the plank under the centre of the stairs.'

Before fitted stair carpets became the norm, a loose carpet would be laid up the middle of the staircase, held in place with a series of stair rods, usually of brass, attached to the stairs with brackets. When buying a carpet, quality was and is a major consideration. Our forebears, without access to hard-wearing modern yarns, would have chosen wool carpets, although coconut matting was used for dwellings such as country cottages. As to length, *Cassell's Household Guide* recommends that carpets should be 'a yard longer than is needful for each flight of stairs, so that when they are taken up for shaking, "not beating", they may not be put down again in the same creases, and thus at trifling expense the carpets will wear as long again as by the usual method of exact measurement.'

The same guide is stern in its advice over care. 'Much injury,' it says, 'arises to all carpets from servants being allowed to run about with high heeled, and sometimes nailed boots. The only way to get over the difficulty,' it concludes, 'is, when engaging them, to mention that only house slippers can be allowed to be worn by them in the house.'

In the days when all large houses employed servants, rigid distinctions were made between 'above' and 'below stairs', that is, between masters and mistresses and their employees. The expression arises from the fact that from the late 18th century most town houses were built with basements, which the servants occupied during the daytime.

KEEP FOOTWEAR IN SHAPE WITH SHOE TREES

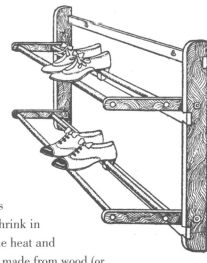

The old-fashioned, but still the ideal way of looking after footwear. The best of all shoe trees are made from cedar wood, which helps to absorb moisture and remove noxious odours from leather and suede.

Shoe trees are particularly useful for footwear that is worn only occasionally, since boots and shoes can shrink in size if the leather is not kept warm and supple by the heat and perspiration of the body. Inexpensive shoe trees are made from wood (or plastic) and metal, but what is most important is that their shape should match the style of the shoe – putting a pointed tree into a round-toed shoe will stretch it out of shape. Bespoke shoes come with all-wood trees, made in several pieces, that fit them exactly. In the absence of shoe trees there are other means of keeping shoes or boots in shape, especially if they are wet when taken off. Scrumpled newspaper is a good solution and, in the autumn, conkers will effectively absorb excess moisture. Another old remedy is to pack the toes tightly with dry sawdust before putting in shoe trees; the sawdust can afterwards be re-dried in a warm oven and reused. What is important is for shoes to be allowed to dry out slowly, away from a direct source of heat, so that they do not become stiff and marked.

In some parts of Britain and America, so-called 'shoe trees' are festooned with footwear. This custom is thought to be particularly popular with women wanting to get pregnant, but shoe trees are also believed to be associated with witchcraft.

SHOE CARE

Some old-fashioned tips for looking after footwear:

If shoes become cracked and dry, rub in plenty of petroleum jelly, then buff them with a dry cloth.

Make shoes waterproof with a warmed mixture of beeswax and suet.

Never scrape off dirt and mud with a knife.

Clean patent leather boots with milk.

When cleaning new boots, rub them with a cut lemon before polishing to get a high shine.

If shoes are too loose, sew a piece of garter elastic on the inside of the back of the shoe.

BRUSHING CLOTHES IS A SIMPLE BUT VERY NECESSARY OPERATION

One of the many mantras of Mrs Beeton, and recommended by her as an essential part of the duties of a valet or footman. The clothes brush remains a vital tool for any well-cared-for wardrobe.

Mrs Beeton arms her readers with considerable detail: 'Fine cloths require to be brushed lightly, and with rather a soft brush, except where mud is to be removed, when a hard one is necessary, being previously beaten lightly to dislodge the dirt. Lay the garment on a table, and brush in the direction of the nap. Having brushed it properly, turn the sleeves back to the collar, so that the folds may come at the elbow-joints; next turn the lappels [*sic*] or sides back over the folded sleeves; then lay the skirts over level with the collar, so that the crease may fall about the centre, and double half over the other, so as the fold comes in the centre of the back.' If necessary, clothes would be folded according to the space in which they were to be stored.

It has been said denigratingly of critics that they are 'like the brushers of noblemen's clothes'.

When choosing a clothes brush, fine, springy bristles are best – a handle is optional. For tougher jobs, an additional brush of stiffer bristle is also useful. So as to be ready for use when leaving the house, clothes brushes were traditionally hung on a hall stand, where men and women could also check their looks in a mirror. The well-dressed man might also have a separate, softer brush for his hats.

Brushing alone might not be enough. An old remedy for removing the shine from a worn black coat was to rub it with fine sandpaper. Another was to boil ivy leaves in water, then strain the liquid, leave it to cool, and sponge it on the offending areas.

RAISE THE PILE OF VELVET WITH STEAM

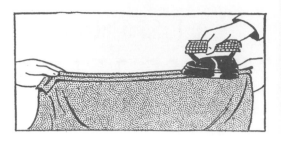

A good way of treating velvet, if done with care. Above all you must avoid pressing it with an iron, which risks leaving indelible marks on the fabric. Deliberate pressing during manufacture, however, creates the beautiful effect of crushed velvet.

Velvet is a luxury fabric that has been made since at least 2000 BC, and the ancient Egyptians documented a technique very similar to one that is still used today. The Italian cities of Lucca, Genoa, Venice and Florence were all centres of velvet making from the 15th century. The fabric is made on special looms that use two sets of warp yarns, one of which may be cut. Royalty and nobility of old would have clothed themselves and decorated their homes with silk velvet dyed in rich colours.

If velvet is creased, it can be steamed or ironed on a velvet board – a fat piece of wood with hundreds of small wires sticking out of it to protect the pile. An old way of dry cleaning velvet was to dip a cloth in powdered magnesia (magnesium oxide powder) and to rub it on to the fabric. Today commercial dry cleaning is a much safer option. For the removal of grease stains, *Enquire Within* recommends the following, somewhat drastic, treatment: 'Pour some turpentine over the part that is greasy; rub it until quite dry with a piece of flannel; if the grease be not quite removed repeat the application, and when done, brush the part well, and hang up the garment in the open air to take away the smell.'

> 'The little old man in black velvet' is an old country description of the mole.

When sewing velvet it is essential to make sure that all the pieces of fabric lie in the same direction. This is because velvet has a 'nap', which affects the colour depending on the effect of light shining on it.

FOLD SILK CLOTHES IN BROWN PAPER

Advice dating back to the days when the bleach used to whiten paper could leach out and impair the colour of precious clothes. Despite the plethora of modern yarn blends and synthetics, pure silk remains a true luxury.

The Practical Housewife of 1860 concurs with the advisability of using brown paper, adding that blue paper is also acceptable and that 'the yellowish smooth India paper is best of all. Silk intended for a dress,' it adds, 'should not be kept in the house long before it is made up, as lying in the folds will have a tendency to impair its durability by causing it to cut or split …'

Silk needs to be treated with care, since it is liable to disintegrate when overexposed to sunlight and is weakened when allowed to get wet, which makes dry cleaning the best treatment. For cleaning silks at home *The Practical Housewife* recommends this procedure, which appears harsh when compared with modern practice: '1. Mix sifted stale bread crumbs with powdered blue [a laundry whitener] and rub it thoroughly all over, then shake it well, and dust it with clean soft cloths … 2. Pass them through a solution of fine hard soap, at a hand heat, drawing them through the hand. Rinse in lukewarm water, dry and finish by pinning out. Finish by dipping a sponge into a size, made by boiling isinglass in water, and rub the wrong side. Rinse out a second time, and dry near a fire, or in a warm room.'

Legend has it that in around 6000 BC the wife of the Chinese Emperor Xi Ling-Shi discovered the cocoons of the silkworms that were eating the leaves of her husband's mulberry trees. That each cocoon is made of a single delicate thread 600–900m (2,000–3,000ft) long was revealed to the empress only when one of them fell into some hot water and unravelled. The Chinese guarded the secrets of their silk making for many centuries – silk fabric did not reach Europe until about

500 BC, when the first silkworms were taken to Constantinople. During the years of Moorish rule in Europe in the eighth century, the rearing of silkworms spread to Spain and Sicily, but it did not reach France for another 500 years. Only in the late 17th century, thanks to exiled Huguenots, did silk weaving spread from France to Germany and Britain.

Parachute silk was especially valued during World War II when such luxuries were unavailable. The misfortune of pilots was turned into triumph, as women seized the fabric and used it to make everything from petticoats to wedding dresses.

KNOW YOUR SILKS

Silk comes in a variety of forms:

Raw silk is thread coated with the natural gum secreted by the silkworms as they spin. It is less smooth than other types of silk.

Wild silk is produced not by the 'regular' silkworm but by the tussah silkworm.

Reeled silk is made by combining several silk filaments from different cocoons. These may be combined to make twisted silk.

Weighted silk is given extra body by the addition of metallic salts. These may, however, impair its durability.

Pure dye silk contains no extra weighting, is strong and naturally elastic.

Satin is silk woven so as to produce a glossy effect on one side of the fabric and a matt surface on the reverse.

TURN WORN SHEETS SIDES TO MIDDLE

Our grandmothers' way of extending the life of bedlinen that is beginning to wear in the middle and a good recession-proof tip. For comfort, a flat seam is essential.

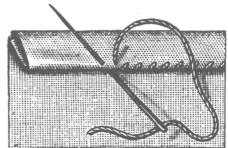

The type of seam best used for the centre of a sheet is a stitch and fell seam. After the sheet has been cut in two down the middle, the two unworn edges to be joined are placed with right sides facing and one edge about 12mm (½in) inside the other. The two sides are stitched together, then the wider seam allowance is folded in and turned down over the seam, and stitched in place by hand or machine. The raw edges at the sides of the turned sheet are hemmed in the usual way.

In the days before the duvet, flat sheets were used on beds, top and bottom, with blankets and perhaps an eiderdown. For warmth in winter and coolness in summer, both linen and silk have long been prized as materials for sheets, being light and absorbent, though cotton is also comfortable.

Top sheets are customarily decorated with an embroidered edging. In her 1888 book *From Kitchen to Garret*, subtitled 'hints for young householders', J.E. Panton advises that '… the top sheet of each pair should be frilled with a Cash's patent frilling two inches and a half wide, and should have a red monogram in the centre to look really well.' The mistress of the house would generally have worked the monograms herself, adding initials to her pillowcases as well.

It is an old superstition that if you change the sheets of a bed on a Friday the Devil will take hold of your dreams for a week. It is also unlucky to sneeze while bed making. If you leave the task before it is completed then the occupant will fail to sleep – or worse.

RESTORE AN OIL PAINTING WITH AN ONION

An old method still recommended today, and one commonly used in times past, as was the application of potato juice. Both the paintings themselves and their frames need good care to keep them in tip-top condition.

An onion cut in half and rubbed over an oil painting will clean it lightly, and the effect can be enhanced by adding a few drops of lemon juice. A wipe over with a damp cloth wrung out in warm water will finish the job. For medium cleaning, experts recommend a lemon detergent in hand-hot water put on with a barely damp sponge rinsed at regular intervals. An alternative is to use a pinch of washing soda crystals in warm water and to apply it with cotton wool, using a circular motion. Again, rinsing with water is necessary.

To say that someone is 'no oil painting' is to be disparaging about their looks.

Although an oil painting may look good hung over an open fire, this is not the best location for it, not least because the heat from the fire will make the painting wrinkle, and smoke and dust will quickly dirty it. As for suitable subjects for this situation, *The Lady's Everyday Book* of the 1870s says that pictures on the walls of a drawing room should depict those 'that awaken our admiration, reverence or love' and that 'prevent our going astray by their silent monitions'. Most favoured in this period were virtuous classical scenes and inspiring landscapes.

In 2008 oil paintings dating to around 650 AD were excavated in Afghanistan. Oil paint (possibly made from walnut and poppy seed oils) was used by the ancient Greeks and Romans, but it was not until the Renaissance that it took over almost totally from tempera – a medium in which the pigments are mixed with egg yolk. The introduction of the modern technique of oil painting can be attributed to the Dutch painter Jan van Eyck, who in around 1410 made paint by boiling pigments, with the additions of ground glass and bones, in linseed oil. Leonardo da Vinci took the process further by adding beeswax, which prevented the colours from darkening.

Picture frames need to be kept well dusted. Wooden ones may be wiped with a damp cloth but not those made of gilt. For the latter, fly marks may be removed with a little vinegar.

THE GREATEST ENEMIES TO FLEAS ARE LIGHT, VENTILATION AND CLEANLINESS

Notorious in history as carriers of plague, fleas are still common pests, flourishing in warm, humid, centrally heated homes. Our beloved household pets can also harbour fleas in their fur.

Fleas are tiny bloodsucking parasites covered with bristles and combs that allow them to cling tightly to their warm-blooded hosts. Their flat bodies are equipped with a powerful pair of hind legs, which enable them to jump from host to host or out of danger. It is a fact that a flea can jump 150 times its own body length and 80 times its own height.

Today, most attacks on humans come from the cat flea, but assaults from the dog flea and the human flea are also possible. Flea 'bites' cause irritation – and sometimes severe allergy – from an adverse reaction to the creature's saliva, which it injects into its host's skin to stop blood from coagulating as it feeds. To prevent flea infestations in the home, regular cleaning, particularly vacuuming, is essential,

> *The way of the world, flea style:*
> *'Big fleas have little fleas,*
> *Upon their backs to bite 'em,*
> *And little fleas have lesser fleas,*
> *and so on, ad infinitum.'*

plus regular and thorough washing of bedlinen as well as pet bedding. Modern anti-flea preparations are widely available and effective.

Under the heading 'Household Hints' *The Girl's Own Paper* for 16 May 1896 records: 'To rid a house of fleas, etc, quassia chips are recommended. Buy one pound from a chemist, put into a gallon of boiling water, and with half this in a bucket of water begin to scour the floors thoroughly. The bitterness of this concoction is a great check … Broken pieces of camphor sewn up in small bags of coarse muslin might be worn in the pocket and put between the blankets and pillows of the bed.'

Bubonic plague resulted from the transference by fleas of a deadly bacillus from rats to humans. Most infamous is the Black Death, which circled around Europe from 1346 to 1665, killing an estimated 75 million people.

KILL COCKROACHES WITH CUCUMBER

Hardly the most effective way of dealing with these ghastly household pests, but at least 'organic'. Apart from preventive measures, traps known as 'Vegas roach traps', filled with coffee grounds, have proved to be one of the best means of exterminating them.

'A good remedy against the cockroach,' says *The Concise Household Encyclopedia* of the 1930s, 'is to strew its nocturnal haunts with fresh slices of cucumber which, being consumed, renders them helpless.' Since this was an age when poisons were commonly used in the home, it is no surprise that the encyclopedia entry continues:

'… a more deadly device is the spreading of phosphor paste on thin bread or mixing it with honey and laying it in their way. Arsenic added to potato boiled and mashed, or mixed with the pulp of a roasted apple is a certain remedy.' Wisely it adds: 'Great caution must be observed in the use of these poisons, or food may get polluted with them.'

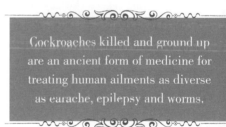

Cockroaches are close relatives of grasshoppers and crickets, but they look much more like beetles and are often wrongly dubbed 'black beetles'. Indoors, they lie hidden during the day and come out at night to feast on food remains – including pet food – and have even been known to bite through book covers to munch on the paste holding the boards in place. They are extremely difficult to eradicate from the home because they not only live in small crevices – for instance beneath and between kitchen cabinets – but can survive for up to three months without food and even for a month without water. This makes the prevention of an infestation, particularly by means of kitchen cleanliness, the best of all remedies.

The Vegas trap works very simply. A glass jar containing coffee grounds and water is placed against a wall. The roaches are able to climb up the wall and into the jar, but cannot escape as they are unable to get a grip on the glass. They can then be squashed or destroyed by some other means. It's said that only cockroaches are attracted by the smell of the coffee, which they love.

Cockroaches killed and ground up are an ancient form of medicine for treating human ailments as diverse as earache, epilepsy and worms.

In some countries, including Russia and France, cockroaches are believed to keep a house safe by acting as its guardian spirits. According to Irish lore, however, cockroaches should always be killed. The reasons are twofold. By some they are regarded as the embodiment of witches; for others they are believed to have been the creatures that betrayed the hiding place of Christ before his death.

MOTH IN CARPET IS A DREADFUL PEST

And will reduce it to threads if left unchecked. Moth is in fact an appalling pest anywhere where it can eat its way through furnishings or clothes.

The clothes moth is a menace. It lays its eggs on the fibres of carpets and clothes, where they hatch into 'maggots' about 6mm (¼in) long; these then plump themselves up by munching holes in your prized furnishings and garments. Wool is their favourite food, but they will also devour cotton and even leather and fur. The larval stage can last for anything between one month and two years, depending on the availability of food. The moths that eventually hatch from the larvae don't eat at all, but are immediately ready to mate and restart the cycle.

In northern England clothes moths are sometimes known as 'ghosts'. Killing one is thought to precipitate the death of a relative.

The 1920s housewife was advised to get rid of carpet moths by scrubbing the floor with '… hot water made exceedingly salty before laying the carpet, and sprinkle the carpet once a week before sweeping until the pests disappear.' Weekly beating of rugs – a strenuous part of the household duties now replaced by vacuuming – was a standard cleaning routine.

Lavender will keep your drawers and wardrobes smelling sweet, but as a moth deterrent it is not nearly as effective as some traditionalists would have you believe. For a natural choice, cedar oil is a better remedy, but still far from totally reliable. Another natural deterrent is allspice berries, which can be sprinkled between clothes in drawers or hung in bags in the wardrobe. Camphor oil was the original ingredient of mothballs, but has been superseded by more powerful and effective chemicals, such as naphthalene and parachlorobenzene. If you use them, follow any instructions to the letter, as these chemicals are highly toxic.

Cleanliness was, and is, the most effective way of protecting items from moths, which are particularly attracted to food and grease on clothes. Keeping woollens well brushed (skin flakes are additional food) will help them stay clean and moth free, and ironing may kill the eggs. Dry cleaning will help to mothproof clothes. Or try putting items in the freezer in a plastic bag for ten days.

FEATHER PILLOWS NEED TO BE WELL SHAKEN

… to keep them fresh and well aired, as befits this luxury form of bedding. Feather mattresses, now a rarity, would once also have demanded the care we now afford to our duvets.

The feather pillow, stuffed with the down of goose or duck, was traditionally covered in a durable case made of a strong, striped linen material called bed ticking. To prevent the sharp ends of the feathers working through the fabric its inner surface would be waxed. 'At certain intervals,' says *Household Science* of 1889, bed-ticking 'should be opened and emptied, the coverings washed and re-waxed and the feathers purified before being put back again. A cleanly housewife,' it adds, 'usually has double coverings to beds, pillows and bolsters, so that the outside ones, which may be of unbleached calico, may be slipped off and washed frequently.'

A feather mattress would have been placed on top of a spring bedstead, in which the metal springs were covered with wool padding and a ticking cover. In very cold weather a second mattress might be used as a bedcover, as we would use a duvet, although an eiderdown would have been more usual.

To be feather-bedded is to be pampered and spoilt. In factories and offices, feather-bedding is the practice of duplicating work and employing over-complicated methods to prevent redundancies.

The duvet, or continental quilt, originated in rural Europe and was usually stuffed with the down of the eider duck. The traveller Thomas Nugent, writing in *The Grand Tour* in 1749, wrote of his surprise on observing the use of the duvet: 'There is one thing very peculiar to them, that they do not cover themselves with bed-cloathes [*sic*], but lay one feather-bed over, and another under. This is comfortable enough in winter, but how they can bear feather-beds over them in summer, as is generally practised, I cannot conceive.'

To test the quality of down in a pillow, press it with your hand. The pillow should spring back quickly, so that your hand does not leave a dent.

SHARPEN KNIVES ON A STEEL

For slicing, chopping and carving sharp knives are a must, because a blunt knife will tear food, not cut it. And because a blunt knife can slip, a sharp knife is, in fact, a safe knife.

When a knife becomes blunt the edge of the metal begins to fold over; the purpose of the steel is to restore that edge to its proper condition. A basic steel is a cylinder of ridged metal, which works by realigning, at a microscopic level, the metal at the edge of the knife. Its advantage is that it can be used regularly without wearing away the knife blade. Modern steels are often coated with diamond, which works by shaving tiny amounts off the knife. With just a few strokes they can make a knife razor sharp. A good steel will have a metal guard at the top of the handle to protect the hand.

On the importance of having the knife good and sharp before carving, *Enquire Within* advises that it should always be 'put in edge' before dinner. 'Nothing irritates a good carver, or perplexes a bad one more,' it says, 'than a knife which refuses to perform its office; and there is nothing more annoying to the company than to see the carving-knife gliding to and fro over the steel while the dinner is getting cold, and their appetites are being exhausted by the delay.'

Cam Weigmann, manager of Henry Westpfal & Company, a 124-year-old Manhattan firm that sharpens cutlery, scissors and leather-making tools, has a golden rule: if you can't cut a tomato, you know your knife needs sharpening.

When using a steel the most important thing to remember is to get the angle right. There are two ways of doing this:

METHOD ONE:
1. Hold the steel with its point on a non-slip surface. Hold the widest part of the knife blade against it at an angle of 45°.
2. Take the knife down the steel, keeping the angle constant.
3. By the time the knife reaches the bottom of the steel, the point should be against the steel.
4. Turn the knife over and reverse the movement, drawing it up. Repeat until the blade is sharp – use 5–10 strokes each side.

METHOD TWO:
1. Hold the steel in one hand and the knife in the other. Cross them, with the wide part of the knife blade at the base of the steel positioned at a 45° angle. Tuck your elbows in.
2. Raise your elbows and part your hands so that the blade moves up the steel, finishing with the tip near the end of the steel.
3. Turn the knife over and repeat for the other side.
4. Repeat until the blade is sharp.

STARCH NET CURTAINS WITH RICE

Saving the water in which rice has been boiled, and which contains grains of starch, makes a mild warm-water treatment for curtains that will not leave them too stiff. Less bother, but more costly, is spray-on starch applied before and during ironing.

Net curtains, the epitome of suburbia, became necessary for privacy long before the suburban estates of the 1930s. Muslin curtains were regarded as essential by 19th-century housewives to protect rooms from the inquisitive gaze of passers by.

Although most starches are now made from cornflour, rice and potatoes were the original stiffening ingredients. The efficient housewife of the past needed to know when to use hot starch and when cold. Hot- or warm-water starch was used for muslins, linens and cottons, including pillowcases; cold starch was for the more pronounced stiffening of collars, cuffs and shirt fronts. Both sorts started with a paste made by mixing the starch powder with cold water to prevent lumpiness. Boiling or cold water was then added, as appropriate.

Mrs Beeton was adamant that 'too much care' could not be given to starched items. 'If the article is lace,' she says, 'clap it between the hands a few times, which will assist to clear it; then have ready laid out on the table a large clean towel or cloth; shake out the starched things, lay them on the cloth, and roll it up tightly, and let it remain for three or four hours, when the things will be ready to iron.'

In the old-fashioned laundry routine, items were dipped in starch after blueing, a practice started by the Romans and brought to Britain from Holland in the 1500s, which worked because the blue rinse masked any yellowness in white fabric.

Originally made from powdered crystals of cobalt blue or lapis lazuli added to the rinsing water in a flannel bag, laundry blue was not superseded until the invention of optical whiteners – colourless 'smart' dyes, which absorb ultraviolet radiation and re-emit it as visible light – in the 1940s.

Borax, or 'laundry booster' was often added to both hot and cold starch to improve results. It works, essentially, as a water softener.

LOOSEN A GLASS STOPPER WITH OLIVE OIL

An effective method of dealing with a stopper that is performing its intended task of making a bottle or decanter airtight but which refuses to budge. The oil works by lubricating the stopper's ground glass surface.

Warmth is also needed to make this method work, as *Enquire Within* explains: 'With a feather, rub a drop or two of salad oil round the stopper, close to the mother of the bottle or decanter, which must then be placed before the fire, at the distance of about eighteen inches; the heat will cause the oil to insinuate itself between the stopper and the neck. When the bottle has grown warm, gently strike the stopper on one side, and then on the other, with any light wooden instrument; then try it in the hand: if it will not yet move, place it before the fire, adding another drop of oil. After a while strike again as before …'

Before the advent of metal caps on bottles in the 19th century, cork and glass stoppers were the norm. As well as bottles for drinks, ground glass stoppers, made from the 1730s onwards, were used to seal scent bottles and those containing smelling salts and medicines. Many old scent bottles, such as those made by René Lalique in the 1920s, are now collectors' items worth £5,000 and more. Also collectable are double-ended bottles, popular from the early Victorian era, designed to hold scent in one end and smelling salts in the other.

The original decanter was the amphora of ancient Rome. The modern version owes its origins to the Venetians, who introduced them during the Renaissance and pioneered the style of a long slender neck opening to a wide body. Its advantage is that it increases the surface area of the wine, allowing it to react with air to improve its taste.

To shift a reluctant screw top from a bottle, try wearing rubber gloves, or grip it with a piece of sandpaper. Or put the bottle under the hot tap: the metal cap will expand and should loosen.

PRESSING IS ESSENTIAL TO GOOD DRESSMAKING

Because this is the best way to make sure that seams lie flat and that the finished garment looks as professional as possible. Other good advice for the dressmaker is to press light fabrics lightly and heavy ones heavily.

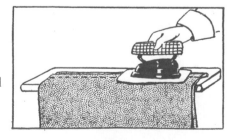

The first heated domestic irons, which were filled with hot coals, were used in the Far East in the eighth century BC, and box irons of a similar design were used until the 17th century. Solid flatirons, which could be heated directly on the stove, were then introduced. Keeping irons hot demanded constant vigilance. If the heat of the stove was allowed to slacken then the irons would be too cool, but the wise housemaid would have 'several irons in the fire' to cover every eventuality. The electric iron was first patented in America by Henry Seely in 1882.

The ironing board is said to have been 'invented' by the Vikings, who spread their clothes on whalebone plaques and smoothed them with wooden rollers, but even well into the 20th century many housewives did their ironing on a kitchen table padded with folded sheets or towels.

The sleeve board is a most useful item, but it is now little seen and is hard to buy. An old rolling pin put into a loosely fitting cloth bag stuffed with old fabric makes a good home-made substitute.

RULES FOR IRONING

Instructions set out in *The Housewife's Handy Book* of 1933, before the introduction of the steam iron, remain useful today:

1. Always see that the iron and all cloths are clean.

2. Be sure to test the iron's heat on a piece of spare material before beginning to press.

3. Never press directly on the material at first. Always have as an intermediary, either a piece of the same material, a piece of muslin or soft cambric, or, in the case of silks, etc., tissue paper.

4. Don't use water [steam] unless you have first tested it on a spare piece of material to see if it marks.

5. Press all cloths with a nap the same way as the nap and not against it.

6. Do not push the iron along the seam but lift it, put it down with pressure and put it down again a little further.

AN OYSTER SHELL IN A KETTLE WILL PREVENT SCALE

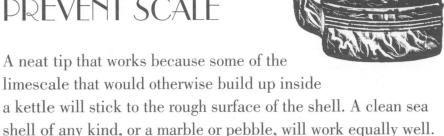

A neat tip that works because some of the
limescale that would otherwise build up inside
a kettle will stick to the rough surface of the shell. A clean sea
shell of any kind, or a marble or pebble, will work equally well.

Limescale is essentially the chemical calcium carbonate, which, when hard
mineral-rich water is heated, is deposited inside kettles and hot water pipes within
the home as a hard, scaly coating. It acts as an insulator around the element or on
the base of the kettle and will reduce its efficiency if allowed to build up.

If you do use an oyster shell – or something similar – in your kettle, it needs to
be renewed every few months; even so it is unlikely to be able to prevent the
problem completely.

If scale is a problem, a simple way of getting rid of it is by soaking with
vinegar: this, being acid, reacts with the alkaline limescale to dissolve it. To treat
heavy scaling, fill the kettle with a mixture
of malt or strong white pickling vinegar,
dissolved half and half with water, and leave
it overnight before rinsing the kettle out
several times with clean water. If there is a
lot of limescale you may need to repeat the
operation. Soaking a clogged-up shower head
in a similar mixture also works well. Another
way of descaling a kettle is to pour in a can
or two of cola. Boil the kettle, leave it to
stand until cool, then rinse it well.

*Oysters were a favourite food of
the ancient Greeks who served
them as an incentive to drink.
The Romans imported them from
England, kept them in salt water
pools, and fattened them up by
feeding them wine and pastries.*

Get scratches off wood with a walnut

Worth trying, as is a mixture of oil and lemon juice, but a proprietary product may give you a better result. For water marks, mayonnaise is an old-fashioned – but by no means infallible – remedy.

The idea behind all of these treatments is that the oil they contain will improve the condition of damaged wood. A walnut will also impart a little stain, so helping to disguise the problem. For water marks, you will need to gently rub the affected area for half an hour or more, so that the oil can work into the wood, or leave the mixture in place for 12 hours before buffing; even then you may still have a problem. Other old remedies are ash and toothpaste, but because these are abrasive they can ruin the finish of a piece. A final, thorough waxing may help, but the bottom line is that sharp objects and water or wine in any form are bad for wood, and a valuable piece may need professional restoration.

Heat can also damage furniture. Placing a veneered piece next to a radiator can draw the natural moisture from the wood, causing the surface to lift and cockle.

Protect tables with trivets, tablecloths and coasters and make sure that lamp bases are protected with felt. Coasters were invented in the 1760s to avoid wine spillages. They were not flat like modern table coasters, but like shallow dishes. They get their name from the after-dinner custom of rolling back the tablecloth and 'coasting' (sliding) the port, in its smooth-bottomed container, from drinker to drinker.

Renovate an umbrella with strong tea

This is hardly the most effective treatment, but one used in past times when umbrellas were more valued than they are today and when prudent umbrella owners would effect their own repairs. No respectable Victorian, or even Edwardian, hallway would be without an umbrella stand.

Umbrella owners of the 1930s were advised that half a cupful of strong tea and two tablespoons of sugar, sponged liberally on to an opened umbrella, would work because the tea would revive the colour and the sugar would stiffen the fabric. When umbrellas were first used against the rain – rather than for protection against the sun (their original purpose) – the silk, linen or cotton of which they were made was oiled or varnished, but it was then discovered that when the untreated fabric was tightly stretched over the metal frame it repelled the rain of its own accord.

Early umbrellas had ribs of wood or whalebone, but by the end of the 18th century metal ribs were in use in France. Heavy umbrellas were nicknamed 'Robinsons' from Robinson Crusoe, the castaway who, in Daniel Defoe's novel of 1719, made an umbrella of animal skins with 'the hair outwards, so that it cast off the rain like a pent-house'. The real revolution came with the foldable steel-ribbed umbrella, invented in 1852 by Samuel Fox and made in his Stocksbridge Steelworks near Sheffield.

If caught in the rain, the Londoner of the mid- and late 19th century could rent an umbrella from one of the many depots of the London Umbrella Company for a deposit of four shillings and a hire rate of fourpence for three hours, or ninepence for the day, with an additional sixpence for a subsequent 24-hour period.

UMBRELLA CARE

'An umbrella when properly treated,' says *Cassell's Household Guide*, 'will last twice as long as one that is not so used.' Recommended precautions, many not necessary for today's umbrellas of stainless steel and synthetic fabric, included:

When wet, do not distend it to dry, which will strain the ribs and covering.

Do not roll a wet umbrella, as this will rot the fabric and rust the ribs.

Do not throw a silk umbrella carelessly into an umbrella stand or the fabric will crease.

When out walking, always hold the umbrella by its handle or you may mark the silk of its covering.

When not in use, protect an umbrella with its silk or oilskin case.

Clean a silk umbrella with a clean sponge and cold water; brush an alpaca one with a clothes brush.

Flatten a curling rug with a damp towel

Certainly a method of dealing with a curling rug, but by no means the only one. It is also essential to make sure that rugs placed on uncarpeted floors do not slip and cause accidents.

If using the damp towel method, it needs to be put in place overnight and weighted down to achieve maximum effect. You must then leave the rug for several hours to dry out before placing furniture on it. An old way of treating a small rug was to stiffen it by painting the back with lacquer then, when it was dry, wiping it over with a weak solution of laundry starch.

> To stop a rug slipping, put a thin rubber mat underneath it.

Rugs come in many guises, from expensive Oriental silks to handmade wool. Rag rugs were once common and remain a good way of using up old textiles. To make them, strips of fabric about 12mm (½in) wide are pulled through holes in a canvas backing with a crochet hook then twisted or tied together to form the 'pile'. Alternatively, the strips of rag can be plaited into braids, which are threaded into the canvas and sewn in place in an attractive pattern.

In homes of old, animal skin rugs were commonplace, made of anything from

sheepskin to tiger, the latter possibly the 'bag' from a hunt in some imperial location. *Enquire Within* says that 'Large rugs of sheepskin, in white, crimson, or black, form comfortable and effective hearth-rugs for a drawing room or dining room. In the winter these may be removed and an ordinary woollen rug laid down as long as fires are kept up.'

A good rug should never be beaten – brushing it then rubbing it on both sides with a cloth wrung out in hot salted water is quite sufficient.

THERE ARE FEW THINGS MORE INJURIOUS TO HEALTH THAN A SMELLING DRAIN

For the logical reason that a smelling drain is one that is most likely to be harbouring the agents of disease. Blocked drains can cause problems if they overflow, so it pays to keep them clear.

When household drains were made of bricks and mortar, it was commonplace, as *Our Homes* relates, for the mortar to become loose. As a result rats could work their way through any holes, 'carrying foul matters, and perhaps the poisons of certain communicable diseases, with them to various parts of the houses …' To add to the problem, drains were often ventilated by the overflow of the lavatory cistern, which not only created foul smells but was almost certainly the cause of typhus infections.

Even house drains made impervious to water needed to be fitted with traps. As *Cassell's Household Guide* explains: 'The sinks, and all water-courses for carrying sewage should be well and carefully trapped, so as to prevent the escape of noxious gases from the sewer into the house. It often happens,' the *Guide* continues, ' that servants carelessly remove the bell-shaped trap from over the escape pipe of the kitchen sink in order to allow of the more rapid escape of the water therefrom, thereby allowing the uninterrupted escape of foul air into the house.'

Great advances came with the manhole, which allowed ventilation, and the U-bend, which prevented reflux. To keep drains clear it is sensible to avoid putting greasy mixtures down them, since the grease coagulates in the pipes. In case of a blockage, pour boiling water or a proprietary drain clearer down the drain.

CHAPTER 3

KITCHEN WISDOM

When it comes to good management, the kitchen remains the boardroom of the home. And it is here that the motto 'waste not, want not' can most readily be fulfilled. In our somewhat straitened times we are becoming aware of the wisdom of our grandmothers, so that the virtues of making a meal from leftovers – the tasty *réchauffés* such as rissoles recommended in the many leaflets produced in wartime Britain – are being extolled once again. Similarly, the virtues of slow-cooked, less expensive cuts of meat, from pork belly to oxtail and shin of beef, are becoming fashionable a second time around. In the economical kitchen, scraps of bread are toasted and made into breadcrumbs, and bones from the carcass of a chicken or turkey boiled to make stock.

The best cooks have always known how to choose their ingredients well, so guarding against wasting money on inferior items. If they are lucky they also have home-grown fruit and vegetables to hand and know how to cook them properly so that they retain their nutritional value as well as looking good on the table. In the plentiful seasons of summer and autumn they will make jams, pickles and chutneys and, unlike their predecessors who spent hours bottling for the winter, stock their freezers in quick time with food to last all winter.

A well-run kitchen is also clean and tidy. Although homes now boast dishwashers in place of the scullery maids of the past, it still pays to wash up as you go and to know how to care for good china, glass and cutlery.

RUB A SUGAR LUMP ON A LEMON TO EXTRACT ITS OIL

A good way of obtaining maximum flavour from the rind of a lemon without getting any of the bitterness of the pith. With hot water added this is a great way of making an 'instant' lemonade. Crushed with a pestle and mortar, 'lemon lumps' also make excellent ingredients for custards, cakes and biscuits.

The merits of the lemon – oil and juice and even the pips – have long been appreciated, but it is likely that lemons were used for medicinal purposes well before their culinary merits were discovered. In ancient times, slices of lemon were chewed to relieve toothache, and lemon peel was used as a contraceptive. In cooking, lemon zest, in which the oil is contained in small spherical glands, adds an intense citrus flavour. More subtle notes come from the oil contained in the sacs or vesicles in the flesh of the fruit.

Sugar cubes were first made in 1840 by the Austrian Jacob Christoph Rad by pressing moistened sugar into sheets, which were then cut up into cubes. A similar process is used today. No polite tea party hostess of the late Victorian era would dream of serving sugar to her guests except with silver sugar tongs (ideally with ends shaped like claws for a perfect grip). Such tongs predate sugar cubes because they were used from the 1790s for picking up pieces broken off loaf sugar – a block of hardened sugar syrup.

In southern Italy, lemon zest is used to make the liqueur limoncello. The most authentic version of all is said to come from Sorrento on the Bay of Naples. It is traditionally served as a digestif.

A TOMATO SAUCE IS A GREAT ECONOMY

And the basis for all manner of dishes, particularly those that come from Italian cuisine. An excellent, good-value sauce can be made with canned tomatoes when, in winter, fresh ones are expensive and often tasteless.

Every cook knows a basic recipe for tomato sauce, and there is nothing quite like the flavour of a sauce made with home-grown, sun-ripened fruit. Onions are essential for a good flavour; otherwise, additional ingredients such as garlic, basil, red wine, peppers and celery are optional. Fresh tomatoes need to be skinned, which is most easily done by plunging them into boiling water to make the skins shrink and split.

The original tomato sauces were probably made by the Aztecs of Central America, who were the first people to cultivate the fruit. In Europe, the tomato was regarded with suspicion as a poisonous 'botanical curiosity' when it was brought from the New World in the 16th century. The oldest known tomato sauce recipe, for 'Spanish tomato sauce' comes from the cookbook of Antonio Latini, published in 1692, and is more akin to a modern salsa: 'Take a half dozen tomatoes that are mature and put them over the coals and turn them until they are charred, then carefully peel off the skin. Cut them up finely with a knife, and add onions finely cut up, at your discretion, finely chopped peppers, a small quantity of thyme or pepperwort. Mix everything together and add salt, oil and vinegar. It will be a very tasty sauce,' he adds, 'for boiled meat or whatever.'

A sieved tomato sauce, boiled for several hours until it is concentrated and well reduced, becomes a tomato paste. Passata, however, is made from chopped and puréed uncooked tomatoes.

SAUCY FAVOURITES

Nothing more than a good tomato sauce is needed to dress a bowl of penne, gnocchi or spaghetti, or to accompany grilled fish or meat. Or it can be used in many other ways:

With meatballs as in the Italian *polpette alla casalinga*.

With peppers, courgettes (zucchini) and aubergines (egg plant) in a ratatouille.

As the basis for a *ragù*; a true Bolognese is a *ragù* of beef and vegetables in a tomato sauce.

With seafood to make a *marinara* sauce for pasta.

THE CREAM OF TODAY IS THE CHEESE OF TOMORROW

As is milk. In the days before refrigerators and the routine pasteurization of cows' milk it turned quickly, and economical housewives would make use of sour cream or milk by transforming them into soft cheese.

The simplest way of making a cheese from sour milk is to wrap it in muslin, hang it up in a cool place and allow it to drip over a bowl until solid. When all the moisture (whey) has drained away the curd is then wrapped in fresh muslin, scalded in boiling water and mashed with salt and pepper before being used. Treating sour cream in the same way produces a cream cheese.

The youngest cheese was once known as green cheese. Gervase Markham, the 17th-century writer on food, also refers to 'Spermyse', which he defines as being made 'with curdes and with the juice of herbes'. It was the custom for so-called milk wives to add herbs of their choosing, as is done with soft cheeses today, although Markham was somewhat dismissive of the practice, concluding that often the herbs clashed, 'not one agreynge with another ...'

For a home-made curd cheese, for use in tarts and cakes, it is necessary to add rennet, a natural substance obtained from a cow's stomach that is also the key ingredient in the making of commercial cheeses. A good 1930s recipe for cheesecakes reads: 'Make a curd by adding a few drops of rennet to 1 pint of hot milk. Press the whey from the curd. Mix with the curd 3 eggs, 3 ratafia biscuits, 4 tablespoons cream, 2oz sugar and a few drops of lemon, pound in a bowl with a wooden spoon, and press the curd in a napkin to absorb moisture. Line six patty pans with puff paste, fill up with the custard and place two strips of candied peel on top of each.' The cakes are then baked in a hot oven for 15 minutes.

Today, rennet is available from specialist suppliers of cheese-making ingredients and equipment, in liquid, tablet and powder form. In times past it was also used for making junket, a delicate 'set' milk pudding, usually flavoured with nutmeg or cinnamon and, for adult palates, a few teaspoons of brandy, and particularly favoured in the Georgian era.

As she sat on her tuffet, before the spider frightened her away, Little Miss Muffet ate her curds and whey together. For full enjoyment they would have been sweetened with sugar.

An inferior joint from a first class animal is better than a more popular cut from a second class one

A tribute to the importance of good quality when it comes to choosing meat. For the very best, choose a good butcher who knows the provenance of his wares or buy your meat direct from the producer at a farmers' market.

> Beating meat to tenderize it has long been practised but is now regarded as less effective than marinading. According to the old rhyme, 'A wife, a steak and a walnut tree – the more you beat 'em the better they be.'

As a rule, the leaner a cut of meat the more it will cost. This relates to its position in the anatomy of the animal and the amount of work the muscles needed to do while the creature was alive. This explains why fillet steak, for example, which is taken from the tenderloin alongside the spine, and has, literally, a supporting role, is so tender and costs many times the price of, say, a piece of brisket taken from the hard-working top of the forelimb.

Less expensive cuts of meat may require longer cooking, but they are full of flavour and can be used to create some superb dishes. Neck of lamb or mutton stewed with onions and topped with sliced potato makes a classic Lancashire hotpot. A breast of lamb can be stuffed and braised or cooked with beans in a rich tomato sauce. Ribs of pork are perfect when covered with a barbecue sauce and roasted. Daubes of beef, cooked long and slow and made with economical cuts such as shin, and oxtail stew, are among the best of all casseroles.

Unless meat is certified organic it is impossible to tell by looking at it whether the animal has been fed a good diet and whether it is free from treatment with antibiotics. However, good meat will always be firm, not flabby, be free of moisture and have no disagreeable smell.

BEEF CLASSIFIED

Mrs Beeton assigned the various cuts of beef to classes according to their quality:

FIRST CLASS – sirloin, with the kidney suet, the rump-steak piece, the fore-rib. The prime cuts, all ideal for grilling or pan frying and roasting. A large piece of sirloin makes one of the best of all roasts.

SECOND CLASS – the buttock, the thick flank, the middle-rib. Use these for braising and casseroles as well as steak and kidney puddings and pies. They can also be successfully slow or pot roasted.

THIRD CLASS – the aitch-bone, the mouse-round, the thin flank, the chuck, the leg-of-mutton

piece, the brisket. Also good for slow cooking. The brisket is traditionally cured in brine before being boiled and served with carrots. Thick flank can be beaten flat and used for beef olives.

FOURTH CLASS – the neck, clod and sticking piece. Cuts that make good soups or, removed from the bone and minced, everything from meat balls to a meat loaf.

FIFTH CLASS – the hock, the shin. Best used for stocks, which can be concentrated so that they will jellify, creating a perfect aspic.

THERE'S A USE FOR EVERY PART OF A PIG BUT THE SQUEAK

A saying that celebrates the value and versatility of the pig, on which, it has been said, the civilizations of Europe and China have been founded. The art of preserving pig meat has a history going back at least two millennia.

Wild pigs were hunted for food by the ancient Britons from before 5000 BC, but these were scrawny creatures, which had little tender meat on their bones. The Celts kept herds of pigs, leaving them to feed on acorns and beech mast in the woods, and this method continued into medieval times in the country, while in towns many families kept a pig or two in a sty, feeding on scraps and domestic rubbish. The improved breeds of pigs we know today began their development in Europe in around 1760, when the innovative Leicestershire stockbreeder Robert Bakewell crossed European pigs with plump Chinese pigs, the latter having been domesticated since the third millennium BC.

In the country it was the tradition to kill a pig in November and to preserve much of the carcass in brine for eating over the winter. Meat was salted to make bacon and hams – air-dried versions such as Parma, Serrano and Bayonne hams are justifiably prized delicacies. Fresh meat, offal and blood was used to make sausages, puddings

The skin of the pig is made into leather, and its bristles are used to make brushes – so extending the animal's usefulness even more.

and other such products. These had been made and appreciated in ancient Babylon, Greece and, especially, in Rome, where pâtés created from the livers of pigs force-fed on dried figs and honey were regarded as a delicacy.

The French art of the *charcutier* began its evolution in medieval Paris where, in 1476, a group of butchers banded together by edict of the king to sell only cooked pork and raw pork fat. By the start of the 17th century the *charcutiers* were also allowed to be purveyors of raw meat and so their range expanded. Their recipes were closely guarded secrets.

MORE DELICACIES FROM THE PIG

All manner of tasty dishes can be made from the meat of the popular porker:

TROTTERS – delicious braised and served hot with a mustard sauce, or cold in a vinaigrette.

PIGS' EARS – a Chinese delicacy when stewed and served with a soya-based sauce.

BELLY PORK – excellent slow roasted.

RILLETTES – a kind of potted pork; shredded meat topped with a delicate lard.

FRANKFURTERS – dry, smoked sausages from Germany.

SUCKLING PIG – roasted and carved from the whole beast.

BRAWN OR GALANTINE – meat from a pig's head set in the jellied stock from the bones.

KIDNEY KEBABS – alternated on skewers with mushrooms and streaky bacon.

A GOOD COOK WILL NEVER HURRY HER ROASTS

If, that is, the meat is to be well cooked all through. Slow roasting allows the fat in the meat to melt gradually into the flesh to give a succulent result. For moistness, regular basting during roasting with the fat that runs from the meat is also recommended.

In an open hearth a spit would rest on iron dogs (stands) placed on either side of the hearth, and be turned by spit-boys. A long dripping pan was placed below the meat to collect the fat that melted from the roast.

The exceptions to the slow roasting ideal might be a joint of beef being cooked so as to be sealed on the outside but rare in the middle, and pork with crackling. For the latter, high heat, either at the beginning or end of roasting (cooks vary as to their preference) is one way of getting maximum crunch.

Before the invention of the oven, meat was roasted over an open fire on a hand-turned spit. In medieval times meat cooked in this way was the preserve of the wealthy and perhaps a treat for lesser mortals on high days and holidays. Even in the Victorian era spit roasting was commonly practised, and *Enquire Within* extols its virtues: 'The Spit has this advantage over the Oven, and especially over the common oven, that the meat retains its own flavour, not having to encounter the evaporation from fifty different dishes, and that the steam from its own substance passes entirely away, leaving the essence of the meat in good condition.'

A VERY OLD FOWL CAN BE MADE AS TENDER AS A CHICKEN

A true fowl is, in the 21st century, a rare bird indeed, but when young tender chickens were beyond most people's means the fowl was a staple of the economical kitchen. To make its tough flesh fit to eat, boiling was the preferred cooking method.

Cassell's Household Guide recommends this treatment for boiled fowls, which was also economical with heat: 'Tie them round with tape, singe and dust with flour, put them in a kettle of cold water in a floured cloth; cover close, set on the fire and take off the scum when it begins to rise; cover again and boil very slowly twenty minutes; take them off, cover close, and the heat of the water will stew them in half an hour.' For serving, it says: 'When taken up, drain, and pour over them white sauce or melted butter. Garnish with parsley and serve with oyster sauce, parsley, butter, or white sauce.'

It is likely that a tough fowl would take longer to cook by another half hour, if not more. It was also common practice to lay slices of lemon over the breast to keep

the meat white. For *poule au riz* rice is added to the pan for the last half hour of cooking, during which time it absorbs the stock, which by then is well concentrated, and the fat from the bird.

Despite its name, a boiling fowl can also be successfully casseroled. A 1937 recipe book for gas cookers suggests jointing the fowl, dipping the pieces in seasoned flour and browning them in dripping with a sliced onion and half a pound of skinned, chopped tomatoes. They are then transferred to a casserole and stock is added to half fill the pot. The fowl is cooked in a slow oven for about three hours or until it is tender and the flesh is falling off the bones.

'It is not desirable,' says the American cookery pioneer Fannie Merritt Farmer – and for obvious reasons – 'to stuff a boiled fowl.'

A boiling hen is a key ingredient of the Italian *bollito misto*, a mixture of boiled meats including osso bucco, beef brisket and a highly flavoured pork sausage such as cotechino.

A FRESH EGG SHOULD FEEL HEAVY IN THE HAND

This is one way of testing an egg for freshness, but by no means the most reliable. A much better way is to put it in water – if it is fresh it will sink. In the days before fresh eggs were available all year round they would be preserved in a brine-based mixture for the winter.

A stale egg floats in water because as the egg ages it accumulates gas, and the more gas it contains the greater the effect. Another test is to hold the egg up to the light. The shell is almost transparent and it will be possible to distinguish the yolk and the white. A stale egg is more translucent at the ends than in the middle, while with a fresh egg the reverse is true.

The quality of eggs is determined by the way in which birds are fed; it is a myth that brown eggs taste better than white since, as a rule, white hens lay white eggs and brown hens brown ones. The colour of the yolk depends on the birds' diet. Those fed on maize or other plants containing large amounts of natural orange pigments have yolks with the deepest colour.

To preserve eggs for the short term an old country method was to rub them all over with melted lard – this was to prevent gas from entering through the pores in the shell – and to layer them in boxes with oats or bran. Alternatively, this was a typical method for 75 eggs, already rubbed with butter as soon as they had been laid: 'Put them into a stone jar with the narrow ends downwards and pour over the following brine: ½ pint slaked lime, ½ pint salt, 1oz cream of tartar and 2 gallons of water. The ingredients should be boiled together for 10 minutes and skimmed. Pour the liquid carefully over the eggs when cold.'

In 1956 Britain's Egg Marketing Board introduced the lion logo, which was stamped on to eggs as a mark of quality. This was combined with an advertising campaign that used a variety of slogans, including the now famous 'Go to work on an egg'. The lion symbol was dropped in 1971 but reintroduced in 1998 for eggs from hens certified as vaccinated against salmonella.

It was long believed that eating too many eggs would 'hard boil' the heart and increase the risk of heart attacks by raising levels of blood cholesterol. Research by the British Heart Foundation has now proved that eating eggs has minimal effect on cholesterol – good news for all egg lovers.

GREASE CAKE TINS WITH BUTTER WRAPPERS

A frugal hint from granny's kitchen, but papers from lard or a brushing of oil are even better for preventing cakes from sticking. For many cakes, lining the tin with lightly oiled greaseproof paper, or baking parchment – which needs no such treatment – is best of all.

The reason why butter is not best for greasing cake tins is that it contains salt, which actually increases stickiness, whatever the method. After greasing a cake tin the cook of old would then roll a little flour around it as an additional precaution against sticking. To line a tin with paper, you need first to cut a strip of paper about 5cm (2in) wider than the height of the tin and long enough to go all round the side, turn up about 2.5cm (1in) at the base and snip it diagonally at intervals. Then cut a round to the exact size of the base. Brush the inside of the tin with a vegetable oil (not olive oil, which will leave a taste) then place the strip inside, overlapping the flaps of the cut base so that it fits snugly. Finally, place the round of paper on top to complete the lining.

When the cake is done, care is needed to remove it in one piece so that it can be left to cool on a rack. 'To remove a cake from the tin without breaking,' says W.H. Steer's *Household Encyclopedia*, 'set the cake-tin on a damp cloth and the moisture created inside the tin loosens the cake until it can be slipped out easily.' Another good tactic is to invert a plate over the cake and turn it upside down. Then peel off the paper, place a cake rack on the cake base and turn it over once again.

Cake making became easier from the early 1900s when greaseproof paper became available, although it was an expensive imported luxury in Britain until the 1930s. Baking parchment, which dates from the 1970s, is a type of wood pulp paper manufactured so as to be impervious to liquids and coated with silicone to give it anti-stick properties. Parchment is ideal for making the perfect Swiss roll or roulade. For a sweet version it can be scattered with caster sugar or cocoa powder or, for a savoury dish, with fine breadcrumbs and herbs.

MORE TIPS FOR SUCCESSFUL CAKES

Every care is needed to ensure a perfect result:

Light sponge cakes need a moderate oven, rich fruit cakes a slow one.

When making a large cake, do not open the oven door for at least 20 minutes – ideally leave it for longer.

Test a cake with a skewer inserted in the centre. If it comes out clean then the cake is done.

Never move a cake in the oven until the centre is thoroughly set.

When cooked, the top of a sponge cake will bounce back when pressed with the finger.

Allow a cake to cool in the tin – it will shrink a little and be easier to handle. Cake is too soft to handle immediately after it comes out of the oven.

For a sponge cake, or any one that will rise a lot, fill the cake tin only half full.

For a rich fruit cake that needs several hours' cooking, tie a double layer of brown paper around the outside of the tin.

If a cake is stuck to its tin, put it on a plate over a pan of boiling water for a few minutes. The tin should then come away easily.

CHOOSE A FLOUR BY SQUEEZING IT IN YOUR HAND

An old way of testing flour for its starchiness. A good baking flour, full of starch, will quickly form a good lump when squeezed. Today's cooks have a huge range of flours to choose from, each formulated for a specific purpose.

The first flours were made some 12,000 years ago by beating grains of wheat between two stones. This early flour was very gritty, and it was only centuries later, with the introduction of mechanical milling in the 18th century, that it became possible to produce the fine, white flour we use today for making cakes and pastries. White flour is, however, less nutritious than its wholemeal counterparts. Even in 1894, *Enquire Within* was instructing its readers that: 'Whole meal bread may be made by any one who possesses a small hand mill that will grind about 20 pounds of wheat at a time. This bread is far more nutritious than ordinary bread made from flour from which the bran has been entirely separated.' Besides promoting health this method also allowed a saving that 'amounts to nearly one-third which would soon cover the cost of the mill.'

Although both plain and self-raising flours are sold as ready to use, in cake, biscuit and bread making best results will be obtained if flour is sifted into a mixture to ensure that every particle is very well separated. Equally, it should be sifted on to a work surface before rolling pastry or pasta dough.

As well as wheat, other grains are used for making flour. Buckwheat flour, which has a nutty taste (and is gluten free), comes from the cereal *Fagopyrum esculentum*, which was probably first grown as a crop in China around 6000 BC; it is widely used to make pancakes. Rye flour, which comes in light, medium or dark varieties, is used to make sourdough breads, and is the principal bread flour of Central and Eastern Europe. The darkest rye flour is the key ingredient of pumpernickel.

If you dream of flour, it is said that you will have a frugal but happy life, while a young woman who dreams that she sees a dusting of flour on herself will be ruled by her husband, and live a life full of pleasant cares. To dream of buying and selling flour, however, is believed to portend hazardous financial speculations.

A SELECTION OF FLOURS

The qualities of a wheat flour depend on its composition and, crucially, on its protein content:

PLAIN OR ALL-PURPOSE FLOUR – a white flour with the bran and wheatgerm removed, often enriched with the addition of vitamins.

SELF-RAISING FLOUR – a plain white flour with baking powder (a raising agent) added.

WHOLEMEAL FLOUR – flour containing all the bran and wheatgerm from the grain.

BREAD FLOUR – a flour containing at least 12 per cent protein and with a high gluten content, so that it holds its shape well when baked. It is also known as 'strong' flour.

PASTA FLOUR – A light white flour, high in protein, which can also be used for pizza and bread making. It is often described as a 'type O' flour.

AN EXCELLENT BREAD CAN BE MADE WITH POTATOES

Indeed it can, although flour is still needed in the recipe. Potato bread, an ideal way of using up left over mashed potato, can be made with or without yeast, as you choose, and formed into rolls or scones as well as loaves.

Recommending potato bread, *Enquire Within* says that it is 'much superior to that made of flour only'. For every three pounds of flour it recommends the addition of a pound of finely boiled mashed and sieved potato. To this is added yeast and the bread is made 'in the usual way'. As to cost, 'taking in the high price of flour, and moderately low price of potatoes, here is a saving of over twenty per cent.,' which, it concludes, 'is surely an object worth attending to by those of limited means.'

Potato bread has many regional names, including fadge, slims and potato farls, tawty or tattie scone in Scotland and tatie bread in Ireland. Apple potato bread, a speciality of Armagh, is a potato bread wrapped, pasty-like, around a sweet filling of apples.

An unleavened version of potato bread, as traditionally made in Ireland, is an excellent addition to any breakfast and can be made without using the oven. To 500g (1lb) of warm peeled, cooked and mashed potato are added 30g (1oz) of butter, a level teaspoon of salt and about 100g (3½oz) of plain flour, or enough to make a stiff dough. This is then rolled out into a round about 1cm (½in) thick and cut into triangular quarters, which are cooked without fat in a hot, dry pan – or on a flat griddle – until golden brown on both sides.

Ideally, serve potato bread when warm, sliced and slathered with butter. Or fry it in the fat remaining in the pan after cooking breakfast bacon. For a variation in flavour, try substituting oatmeal for half the flour in the recipe.

No dinner can be a success unless the vegetables receive proper attention and consideration

Indeed, there are few things less enjoyable than soggy, poorly cooked vegetables, while nutritionally, overcooked vegetables lose many of their valuable vitamins. The golden rule for success is to cook green vegetables quickly and root vegetables slowly.

'If not thoroughly boiled tender, they [vegetables] are very indigestible, and much more troublesome during their residence in the stomach than underdone meats.'
Enquire Within

The best vegetables of all are those that are eaten – cooked or raw – when super fresh, just picked from your own garden or allotment. They will almost certainly need to be washed to get rid of any dirt, and this is always a wise precaution for shop-bought vegetables, which may have been sprayed with insecticide, fungicide or foliar feed. Many vegetables, even beetroot and Brussels sprouts, are delicious raw, shredded and well dressed. Otherwise, green vegetables should be either steamed or 'steam boiled' in a lidded pan with as little water as possible. When cooked they are best served at once – the longer they are left standing the more their vitamin content will diminish. Before serving, a knob of butter and a scattering of chopped parsley or a similar herb will give a great finish to any dish of vegetables. Or you can grate in a little nutmeg, which is an excellent complement to spinach. Root vegetables such as carrots can be fried off in a little butter and sugar to caramelize them, and there are few more delicious vegetable accompaniments than potatoes roasted in goose fat.

WELL PRESENTED

There is no need for vegetables to be plain. These are some classic ways of serving them.

AU GRATIN – in a cheese sauce finished under the grill for a crisp, browned top. Ideal for cauliflower and onions.

BRAISED – cooked in the oven in a well-flavoured stock: good for celery and leeks.

STUFFED – good for all kinds of vegetables, from peppers, tomatoes and marrows to large mushrooms.

SOUFFLÉ – works best with root vegetables such as carrots, celeriac and parsnips, but also good with cauliflower or broccoli.

SCALLOPED – layered with a mixture of breadcrumbs, herbs and butter. Best with moist vegetables such as tomatoes.

FRITTERS – pieces of vegetable coated in batter and fried – in Japanese cuisine called tempura. Works with everything from courgettes (zucchini) – both flowers and fruit – and aubergines (eggplant) to cauliflower.

USE AN APPLE TO KEEP A CAKE MOIST

Putting an apple in your cake tin is indeed a good way of preventing the contents from drying out. Long keeping is a requisite for success with rich cakes such as Christmas cake.

Putting a piece of celery in the bread bin works in a similar way to keep bread fresh; this is a method long recommended by careful home makers.

The keeping qualities of a cake depend on its richness. Cakes that keep least well are light sponges. As a rule, the greater the proportion of butter, eggs and fruit to flour the longer a cake will keep without going stale. A fruit cake will continue to mature in the tin over several weeks, and will be even moister if it is pricked liberally with a skewer all over the base and doused with a generous amount of brandy or liqueur. Parkins, gingerbreads and tea breads will also improve if kept in an airtight tin or wrapped in foil for a few days before they are eaten.

It is an old tradition to keep the top tier of a wedding cake for the christening of a couple's first child. An iced fruit cake will, indeed, keep for several years if air is excluded, but for storing other cakes, such as a rich chocolate, a freezer is essential.

Ripen green tomatoes with a bunch of bananas

Put these two fruits together and the bananas (unless they, too, are green) will give off ethylene gas, which will assist in ripening the tomatoes. Fully ripe tomatoes will have the same effect.

Anyone who grows their own tomatoes outdoors will know the frustration of having a plethora of unripened fruit at the end of the season. Put in a bowl with bananas they should ripen in a few days. For long-term ripening they can be put in a dark drawer in a cool place where, as long as they are sound, they will gradually turn red and can be used over a long period.

In the kitchen, green tomatoes have many uses. They have a firm texture with a pleasantly acidic bite and just a whisper of tomato flavour and can be used in all kinds of dishes including chutneys, relishes and, of course, fried green tomatoes. The last is a traditional dish from the Southern United States, in which tomato slices are dipped in cornmeal and fried in bacon fat.

Green tomato chutney works best if apples are included in the ingredients; for every 1.5kg (3lb) of tomatoes you will need 500g (1lb) of cooking apples. For a green tomato jam, *The Parkinson Cookery Book* of the 1930s gives this interesting recipe: 'Allow 1lb lump sugar and 1 pint water to every 1lb of sliced tomatoes. Allow grated rind and juice of 2 lemons to 3lb jam, or if liked, 2oz bruised ginger to the same quantity of jam. Tie the latter in a cloth and boil in the jam. The sugar and water should be boiled together until thick; slice the tomatoes thinly and add. Boil until tomatoes are quite clear, boiling either lemons or ginger with it. Pour into jars when cooked and make airtight.'

The film *Fried Green Tomatoes at the Whistle Stop Café* was released in 1991. The café was loosely based on the Irondale Café in Irondale, Alabama, birthplace of Fannie Flagg, who wrote both the film script and the original novel on which it is based.

A LITTLE POT IS SOON HOT

A good way of saying that it is important to choose a pot or pan of the correct size if you want to be as economical as possible with cooking fuel. Putting on a lid will also help to conserve heat within the pot.

Cooking pots are the most ancient of kitchen wares. The oldest clay pots yet found are from Japan and date to around 10,000 BC; metal pots were not made for another millennium, when the Chinese began to make wok-shaped pans.

Whatever their size, the best modern pots and pans to use on top of the stove will be made of a metal such as cast iron, stainless steel or copper. All these work equally well in the oven, but for a gentler but even heat earthenware and treated glass are also good choices.

A good heavy-bottomed saucepan will double up as a stewpan and can be simmered on the top of the stove. It can also be used for steaming, or for the long slow boiling of a dish such as a Christmas pudding.

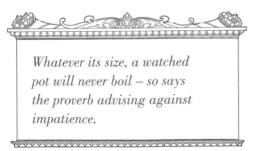

Whatever its size, a watched pot will never boil – so says the proverb advising against impatience.

Extolling the virtues of the saucepan, *Enquire Within* says: 'There are few kinds of meat or fish which the Saucepan will not receive, and dispose of in a satisfactory manner; and few vegetables for which it is not adapted. When rightly used, it is a very economical servant, allowing nothing to be lost; that which escapes from the meat forms broth …' Likening the stewpan to the human stomach, the article maintains that the digestive action of this organ '… is more closely resembled by the process of stewing than by any other of our culinary methods'.

WASH UP AS YOU GO

The ideal way to keep the kitchen tidy as you work, but not always possible if you are serving dishes that require last-minute attention and need to be put on the table immediately. Unlike their predecessors, modern cooks have the boon of the dishwasher to hand.

Even if they cannot be washed up as you cook, it is wise to put any pans that will need serious treatment to soak in hot soapy water as soon as you have finished with them, to make cleaning easier later on.

When washing up the dishes after a meal it is vital to be orderly, and to have everything stacked and plates scraped before you begin. To make best use of your hot water, the method recommended in old-fashioned guides still holds good. This means beginning with glass and silver, then washing china and finally knives and other metal items such as pots and pans.

In warm, dry weather, and as long as the water is not too dirty, you can recycle your washing-up water by putting it on shrubs and other tolerant garden plants.

In the Victorian kitchen there would be different types of cloths for drying different items. So that they might be easily distinguished, the glass cloth would be of a checked linen of medium texture and the china cloth of coarse white linen; thick, unbleached linen towels were kept for knives and metal pans.

The first practicable dishwasher was invented by Josephine Garis Cochrane of Shelbyville, Illinois, and patented in 1886. Mrs Cochrane was a wealthy woman who entertained often, and she wanted a machine that could deal with the dirty dishes faster than her servants could, and without breaking them. First she measured the dishes, then she made wire compartments, each designed to fit plates, cups, or saucers. The compartments were placed inside a wheel that lay flat within a copper boiler. A motor turned the wheel while hot soapy water squirted from the bottom of the boiler and rained down on the dishes. The new machines were expensive, at $150, and required an abundant supply of hot water, so they were not as popular with the public as Mrs Cochrane had expected. The early models were sold mainly to restaurants and hotels, and it was not until the 1950s that dishwashers finally started to become standard household appliances.

HELP WITH THE WASHING UP

More useful tips and hints:

Wash a few things at a time.

Never soak knives with bone handles in water or put them in a dishwasher – you may loosen the glue that fixes them.

Soak pans with burnt food stuck to them in warm water, if necessary overnight.

Rinse off suds before drying – especially from glassware.

Beware of washing up delicate glass or porcelain in very hot water – it may crack.

If hard water has left marks on any items, immerse them in vinegar to make the marks disappear.

A TASTY RÉCHAUFFÉ DEMANDS LITTLE TOIL

But needs considerable care and imagination if the results are to be enjoyable. The merits of the *réchauffé*, well known to frugal cooks of the past, are once again becoming appreciated as an acceptable way of using up leftovers.

A *réchauffé* is what *The Constance Spry Cookery Book*, first published in 1956, calls a 'done up' dish made of any kind of meat, poultry or game, which 'may be heated up in a suitable sauce, dipped in batter and fried, turned into rissoles, croquettes, savoury pancakes, cottage pies and so forth.' Introducing the subject, the book advises that these 'homely dishes' are 'simple to prepare but only if proper care is taken with them.' Cooked fish can also be used.

The French word *réchauffé* means 'reheated', but it is essential to reheat with care. If meat that is already cooked is subjected to prolonged heat it will become tasteless and even tough. The key to a good *réchauffé* often lies in making sure a mixture is moist enough. Adding a sauce or gravy, or vegetables such as chopped tomatoes or cooked mushrooms in their juice, will always be helpful.

As for seasoning, in its tips for using 'What's left in the larder' a World War II instruction leaflet sensibly reminded readers that: 'All dishes using left-overs need more seasoning than when uncooked is used. This is because some of the natural flavour of the food is lost with re-heating.' On a health note, the leaflet recommends that: 'Dishes using cooked vegetables should, as far as possible, be served with a fresh salad, or a serving of freshly-cooked greens to make up for the Vitamin C lost in cooking and reheating.'

NEW DISHES FROM OLD

Some tasty ways with leftovers:

COTTAGE PIE – made with minced beef and vegetables topped with mashed potato. A shepherd's pie is made with lamb.

FISH CAKES – flaked cooked fish with mashed potato and parsley. Dipped in egg and breadcrumbs and fried.

CROQUETTES – finely chopped chicken or other meat mixed with a béchamel and chopped mushrooms. Shaped into small roll shapes, dipped in egg and breadcrumbs and deep fried.

SAVOURY HASH – chopped meat with fried onions and diced potatoes, well seasoned.

BEEF ROLLS – slices of cooked roast beef spread with mustard or horseradish and chutney, rolled up, tied with string and cooked in a thickened gravy with redcurrant jelly and/or port added.

FISH PUDDING – cooked, flaked fish in a cheese sauce, layered with cooked macaroni, topped with breadcrumbs and baked in the oven.

Bits and pieces make a fine stock

Correct, because almost any meat, fish or vegetables can be used to make the stocks on which good cooking depends. In times past, the stock pot was once seen in every kitchen, large and small.

Meat, vegetables and fish are all used to make stocks, needed for everything from soups and sauces to casseroles and risottos. While beef is the basis for a dark meat stock, chicken is more suitable for a light one, but all kinds of meat may be used. When using poultry giblets, however, it is advisable to leave out the liver, which can impart a bitter taste. For a jellied stock, bones – preferably fresh – must always be included. For a fish stock, fish heads and bones, and the shells of prawns and lobsters, can all go into the pot.

Before the arrival of refrigeration the stock pot would be boiled daily to prevent it from harbouring bacteria. Any fat from the pot was skimmed off and used for cooking or spread on bread.

All stocks need to be well flavoured, not only with salt and pepper and with herbs such as parsley, sage, rosemary and thyme, but with hearty vegetables. Stock making is a good opportunity to use up the tough outer stalks of celery, the outer leaves of cabbages and cauliflowers, the stalks of mushrooms, and carrots that are a little past their best. Be careful of adding turnips or parsnips unless you are partial to their taste, as their flavour will dominate any mixture.

Stock cubes owe their existence to the invention of concentrated meat extract by the German chemist Justus Liebig around 1840; it was sold commercially from 1866. In 1899, the company introduced the trademark 'Oxo' for its affordable home version, and the first Oxo cubes went on sale in 1910.

BROWN STALE BREAD IN THE OVEN

A good way of using up bread that is past its best, and an ideal way of making your own breadcrumbs. There are all sorts of other ways of creating delicious dishes using stale bread in some guise.

Bread baked in a slow oven until it is crisp and brown can be crushed into fine crumbs with a rolling pin or, for the finest crumbs of all, pressed through a wire sieve. Stored in an airtight container, dried crumbs will keep for several months and are useful for coating all kinds of foods such as fish cakes and croquettes, and for topping savoury dishes.

Dried breadcrumbs can also be used for sweet dishes, as in this wartime recipe for 'Crumb Fudge', which needed little cooking. The ingredients are: '2 tablespoons syrup, 2oz margarine, 2oz sugar, 2oz cocoa, few drops vanilla, peppermint or orange essence, 4–6oz dried crumbs.' As to the method: 'Heat the syrup, margarine, sugar and cocoa gently until all is melted. Stir in the required flavouring and then the bread crumbs. Mix thoroughly and turn into a well-greased 7in sandwich tin; spread evenly and mark lightly into fingers or squares. Leave for 24 hours and then use as a cake or sweet.'

> On average, more than 325 tonnes of bread are thrown away by British households every year.

Chunks of stale bread baked in the oven until quite dry and hard make good rusks for babies, while smaller pieces are excellent as fat-free croutons. In the Victorian nursery children would sometimes be fed 'brewis', a kind of porridge made by soaking crusts and dry pieces of bread in hot milk, then mashing it up and flavouring it with salt.

ALL KINDS OF FRUIT WILL MAKE FINE FOOLS

They will, but some fruits, such as rhubarb and gooseberries, make the best fools of all. When these confections were first made (probably before the 17th century) some, such as the Norfolk fool, contained no fruit at all, but were simply a rich, spiced egg custard also known as 'white pot'.

An alternative way of making a fool is to add a stiff egg custard to the fruit and cream – made with gooseberries, it is described by Dorothy Hartley in her Food in England *as 'a soft green cloud'.*

The word 'fool' probably comes from the French verb *fouler*, meaning 'to mash', and the simplest fool is just mashed fruit mixed with whipped cream. The practice of cooking and mashing fruit was, thought cooks of the past, a way of making it safe to eat, and a fool would have been a good use for it. The task of whipping the cream would have been accomplished with a bunch of twigs until the fork came into general use in the late 1600s.

There are several notable variations on the fool, including the French *crème printania*, consisting of puréed strawberries mixed with kirsch and crème Chantilly, which is whipped cream sweetened with sugar. In Eton mess, named from the school where it first became popular, mashed strawberries and whipped cream have crushed meringue added. The dish was probably first served in the school sock shop (tuck shop) in the 1930s and was originally made with strawberries – or bananas out of the strawberry season – mixed with either cream or ice cream.

ANY INGREDIENT IS FIT FOR A PIE

A saying that is tribute to the
versatility of dishes covered or encased in pastry. The pie may
get its name from the magpie and refer to the practice of the
good cook who, like the magpie, would collect all kinds of
different ingredients to put in a pie.

In pies made in medieval times the pastry served as an impregnable container for
the meat inside and was not intended to be eaten: at best it was broken up and used
to thicken pottages. The pies of the time could be huge – up to 4m (12ft) wide and
filled with, for example, venison, vegetables, fruit and spices. By Tudor times, pie
pastry had evolved, with the addition of suet, into something more appetizing, and
fruit pies were added to the repertoire. *A Proper New Booke of Cookery* of 1545
includes a recipe for 'short paest for tart' whose ingredients included, as well as fine
flour, 'a curtsey of faire water and a disshe of swete butter and a little saffron and
the yolkes of two egges and make it thik and tender as ye maie.'

   The popularity of pies has never faded, nor their use as receptacles for
everything from steak and kidney to apples, rhubarb or gooseberries. Extolling their
virtues, Alexis Soyer in his *Shilling Cookery for the
People* reminds us that: 'From childhood we eat
pies – from girlhood to boyhood we eat pies –
from middle age to old age we eat pies – in fact pies
in England may be considered as one of our best
companions *du voyage* throughout life.'

   To eat humble pie is to be contrite, but the saying
relates to real food. In the 17th century pie made from
the 'umbles' – the liver, hearts and entrails – was
served to diners at table such as the retainers and
unimportant guests, who thus had to eat humble pie.

> Some of the world's largest pies
> are those made in Denby Dale in
> Yorkshire to commemorate great
> events. The first was made in 1788
> to celebrate the recovery of King
> George III from his 'madness'.
> The most recent was made for the
> millennium in 2000.

SEAL A POT WITH GOOD FAT

A good old way of keeping food airtight to preserve it, essential in the days before refrigerators and freezers. Some modern dishes such as pâtés and potted shrimps are still sealed with a layer of butter.

Mrs Beeton includes a recipe for potted shrimps that has not dated at all apart from the price of the dish, which she calculates to be one shilling and threepence. Her recipe is simple: '1 pint of shelled shrimps, ¼lb of fresh butter, 1 blade of pounded mace, cayenne to taste; when liked, a little nutmeg. *Mode.* Have ready a pint of picked shrimps, and put them, with the other ingredients, into a stewpan. Let them heat gradually in the butter, but do not let it boil. Pour into small pots, and when cold, cover with melted butter, and carefully exclude the air.'

A good fat is one that has a fresh smell with no hint of rancidness. In the days when nothing in the kitchen was wasted – and during World War II when fats were strictly rationed – when any stock pot, casserole or other dish was skimmed of its fat it would be kept and used for frying and other purposes. Similarly, fat from roasted meat was prized as dripping and saved for cooking or spreading on bread.

A quick way of clarifying fat is to slice a large potato, put it into the fat, bring the fat to the boil and simmer it until the liquid is still and the potato dark brown. It can then be strained through muslin. All the impurities will have been absorbed by the potato.

Fat from raw meat such as pork would be rendered down by cutting it into small cubes and heating it very slowly for several hours until it was converted into a clear liquid. This fat was then clarified by putting it into a pan, covering it with water, and simmering it. When it cooled the fat could then be easily removed from the top of the pot.

THE WORK OF PRESERVING REQUIRES TIME AND PAINS

But is well worth the effort, since there is nothing quite like home-made jams and preserves, pickles and chutneys. Cooks of the past also spent much time bottling fruit and even pickled eggs in large quantities.

The key to preserving lies in minimizing the possibility of food being subjected to attack by the agents of decay. In pickles and chutneys, it is salt and vinegar that accomplish this process, while in jams and jellies it is sugar. In bottling, both the heat of the process and the exclusion of air are effective.

'When tomatoes are available,' says the World War II Ministry of Food Leaflet No 24, 'the wise housewife will preserve some for use in the winter. They are valuable then, not only for the colour and flavour they give to dishes, but also for the protective

> For an instant preserve, simply put sound fruit such as cherries into a jar and fill it with brandy.

vitamins they contain.' It then goes on to instruct in detail how to choose and use preserving jars, how to boil and seal the jars and, crucially, how to test the rubber seals that are held in place by a vacuum after the jars have cooled.

The labour of preserving lies in the preparation of the fruit and vegetables, from skinning tomatoes and shallots to chopping peel for marmalade and shredding vegetables for pickles. This is followed by the weighing of ingredients, the stirring and testing until a jam is set or a chutney is thick. Then the preserves must be ladled into sterilized jars, filling them to the top, covered, sealed and labelled with the name and date. Patience is also required to allow time for pickles to mature – it usually takes at least three months for the flavour to develop.

CHAPTER 4

CLEAN AND BRIGHT

Unlike the Victorians, who hung on the words of the Methodist preacher John Wesley, we may no longer believe that cleanliness is next to godliness, but there are still many good reasons for keeping the home – and everything in it – clean and bright. The most crucial of these is, of course, the matter of hygiene. A dirty home is one that harbours the agents of disease and encourages infestation by infection-bearing pests.

Compared with the cluttered rooms of the 19th century, which were filled with nick-nacks, pictures and ornaments of every kind, the modern home is much easier to clean. We also have the advantage of efficient vacuum cleaners and modern detergents, which, together with snugly fitting doors and windows, and central heating instead of open fires, mean that it is no longer necessary to scrub, dust and polish as part of a daily routine. Spring cleaning, however, remains desirable. Every home can benefit from an annual turn out of everything from old clothes to unwanted magazines.

Because every material demands a different method of cleaning. it certainly pays to know how to treat your many possessions, from clothes and jewellery to silver and marble. And many of the old ways and pieces of advice still work well today. Bicarbonate of soda remains an excellent oven cleaner and disperser of grease of all kinds, and the inside of a banana skin will buff up leather. Other treatments, however, are probably best assigned to history, whether it is rubbing treacle into a grass stain or putting white wine on to a red wine stain.

SPRING CLEANING IS A NECESSITY FOR EVERY HOME

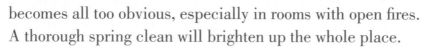

As the days lengthen and the sun rises higher in the sky, the dust and grime that has collected in the home over the winter becomes all too obvious, especially in rooms with open fires. A thorough spring clean will brighten up the whole place.

Drawing up a plan of action before you begin will make spring cleaning most effective and cause least fuss and disturbance to the home. When time is at a premium it is best to tackle one room at a time, clearing out all cupboards and wardrobes as you go. As to the order in which you work, Mrs Beeton maintained that '… it is usual to begin at the top of the house and clean downwards; moving everything out of the room; washing the wainscoting or paint with soft soap and water; pulling down the beds and thoroughly cleansing all joints; scrubbing the floor; beating feather beds, mattresses and paillasses, and thoroughly purifying every article of furniture before it is put back in its place … Warm winter curtains are replaced by the light and cheerful muslin curtains. Carpets are at the same time taken up and beaten …'

As one 1930s encyclopedia pointed out, an electric vacuum cleaner is most useful for spring cleaning. The modern vacuum cleaner owes its invention to such pioneers as the English engineer Hubert Cecil Booth, who created the first suction-based dust-collecting machines (so large that they had to be drawn along the street by horses and parked outside the house during cleaning) and to the American James Murray Spangler, who patented his electric suction sweeper in 1907. Both inventions might have ended up as curiosities in museums had Spangler not given one of his machines to his cousin, Susan Hoover. Her husband, William Henry

Hoover, bought Spangler's patent and in 1926 marketed the first upright vacuum cleaner with a bag connected to an upright handle. It was also fitted with a 'beater bar' and advertised with the slogan 'Beats … as it sweeps … as it cleans.'

The ritual of spring cleaning can be traced back to the ancient Jewish practice of cleansing the home ahead of the festival of Passover, before which all remnants of any leavened food must be removed. Observant Jews will use candlelight to search their houses for crumbs of such *chametz* on the eve of the Passover holiday.

SPRING THOROUGHNESS

Jobs for the careful home owner to attend to as part of the spring cleaning ritual:

Store away winter clothes, protected from moths with cedar chips or balls.

Clean mattresses, blankets, duvets, pillows and other bedlinen.

Clean and renovate carpets and curtains, using professional services if necessary.

Sort out kitchen storecupboards and discard any food items that are past their best.

Wash all china and glass, and clean all ornaments.

Renovate or renew lampshades, cushions and other accessories.

Get rid of all rubbish and surplus items, including clothes: send them for recycling or take them to your local charity shop.

Dust all books and discard unwanted old papers and magazines.

POLISH WINDOWS WITH NEWSPAPER

Rolled into a pad, a good old-fashioned substitute for a polishing cloth or chamois leather, but by no means the only effective way to get the shine on your panes.

Dirty windows not only deprive rooms of light but make them look unattractive. If you can, clean them on a dull day, or at least when the sun is off them, to prevent streaks and allow you to see the results of your labours clearly. Good cleaning choices are either washing with detergent or a proprietary liquid or spray containing ammonia and alcohol, which is then rubbed off with a soft cloth.

'When I'm Cleaning Windows' was a song made popular by the Lancashire singer and ukulele player George Formby and originally recorded on 27 September 1936.

A mixture of equal parts of paraffin and methylated spirits can be rubbed on to dry glass and polished off when dry, and turpentine is especially good for grease marks. Old country ruses include rubbing with a bunch of stinging nettles dipped into water with a dash of vinegar added, or washing with a solution made by pouring boiling water on a chopped-up potato. The squeegee, used by the professionals, is an excellent – and some say the only – way of getting a good finish.

When cleaning, don't forget the frames. Dusting helps to keep them dirt free, but in winter it is also essential to remove any condensation which, as well as rotting wooden frames, can be the breeding ground for unsightly black mildew.

Clean windows were once essential to a good reputation. 'Dirty windows,' it was said, 'speak to the passer-by of the negligence of the inmates.'

CLEAN BOTTLES WITH EGGSHELLS

When marks are inaccessible with a brush, shaking crushed eggshells in a bottle or decanter, with warm water and a little washing up liquid, is an excellent and effective way of cleaning.

Alternatively, you can buy boxes of 'magic' metal balls that do the same job by abrading the inside of the bottle. These are the modern version of the lead shot that butlers traditionally used in decanters, with the addition of brandy. Coal ashes shaken with hot or cold water are another traditional cleaner.

Eggshells have other good uses. Crushed and spread in the garden they can help to keep slugs off your most vulnerable plants. In halves they make excellent containers for planting seeds and nurturing seedlings. In the kitchen they can be used to clarify consommés: the cloudy particles in the liquid cling to the pieces of shell, which can then be removed and discarded.

Crushing or pricking the shell after you have finished eating a boiled egg – or finding some other way of ensuring that there is an aperture at both ends – is said to keep witches or evil fairies away, or to prevent them from changing the shells into boats. But the quickest way to stop hens from laying is, say the superstitious, to throw eggshells into the fire.

The shell of a hen's egg is 95 per cent calcium carbonate and is secreted around the yolk and white by the shell gland, in a process that takes about 14 hours to complete. It is traditional chicken-farming practice to add crushed eggshells to the birds' feed to provide the minerals they need to make fresh shells.

To clean decanters or bottles, Cassell's *Household Guide* recommends this method: 'Tear up soft paper into small pieces; put them into the decanter or water bottle; pour onto them cold water, and shake until clean; then turn them out and rinse.

A PINCH OF BORAX ERADICATES GREASE

Once a mainstay of the household battery of cleaning materials, borax has been largely confined to domestic history since it became an ingredient of many modern detergents. However, it is inexpensive and, being a natural mineral, is in fact environmentally friendly.

A white powder that dissolves easily in water, borax has numerous uses in industry as well as in domestic cleaning. As well as cutting through grease, it has fungicidal and insecticidal properties, and is used in the manufacture of glass and ceramics.

To use its detergent action in your laundry, soak very dirty items such as dishcloths or sponges in a solution of 1 tablespoon of borax in 4 litres (1 gallon) of water. To clean dirty pots and pans, sprinkle on a little of the powder and rub with a damp sponge or cloth. For floors, add about 5 tablespoons of borax, plus a little mild liquid detergent, to a bucket of warm water.

Borax, also known as sodium borate or sodium tetraborate, occurs naturally in various places around the world, including the Mojave Desert in California, Chile and Tibet, where deposits are formed by seasonally evaporating lakes. It is also created synthetically from other boron compounds.

Because, as it acts, it converts some water molecules to hydrogen peroxide (a form of bleach), borax whitens, disinfects and deodorizes as well as cutting through grease. It is a wise precaution to wear rubber gloves while using it, although it is said to smooth skin on the hands and to soften the face.

In the Edwardian home, 1 teaspoon of borax in 500ml (1 pint) of water was used as a healing solution for cuts and wounds. A mixture of 2 teaspoons of honey with 1 teaspoon of borax was prescribed both as a remedy for sore throats and, applied to the skin, for thrush.

OLD-TIME USES FOR BORAX

Some of the many other traditional uses for borax in home and garden:

For cleaning marble if sprinkled on then removed with warm water.

Sprinkled around the home to kill insects and vermin – including cockroaches.

As a disinfectant for closets and sinks and 'wherever sewer gas is suspected'.

Mixed with ammonia to make a shampoo for greasy hair.

In small quantities, for purifying water for cooking and drinking.

Patent Californian borax was sprinkled on meat to preserve it. (This is *certainly not* recommended today.)

Mixed with hot water and sprayed on roses to remove greenfly. Also recommended for wiping down and disinfecting the shelves of greenhouses.

NEVER DRY CLEAN CASHMERE

Gentle hand washing, with a mild liquid soap formulated specially for the purpose, is what the experts recommend. Dry cleaning can harden the delicate fibres of this luxurious yarn, though it helps to kill the grubs of clothes moths.

To wash a cashmere sweater successfully, follow these cardinal rules. First, wash and rinse in water of the same, warm temperature. Do not wring or twist the fabric. Next, spread the garment out on a dry towel, ease it into shape, then roll up the towel. Press gently and repeat with a second dry towel. Then lay the sweater on a third dry towel placed on a flat surface and leave it in the air to dry, but not near a radiator or any other direct source of heat. Finally, fold up the garment with care; do not put it on a hanger.

The Cashmere Lop is a type of long-haired rabbit descended on one side from the English Angora. The soft coat of angora rabbits is used, like cashmere, to make a yarn for weaving and knitting.

Keeping cashmere clean will deter clothes moths, but to prevent their ravages it is wise to store garments in individual polythene bags, with cedar oil added.

True cashmere (not a merino wool substitute) is named from the Kashmir goats that graze in the highlands of Tibet, the Gobi desert and the mountains of Central Asia. For centuries cashmere was enjoyed only by the wealthy, but today it is much more affordable. Napoleon presented his wife Eugenie with 17 cashmere shawls, and in the 19th century the dandy Beau Brummell's white cashmere waistcoats were objects of envy.

GET GREASE MARKS OFF CLOTHES BY IRONING THEM OVER BLOTTING PAPER

The theory is that the heat of the iron will melt the grease – on carpets as well as clothes – which will then be absorbed by the blotting paper. In practice, however, you need to be careful you don't 'cook' the mark and make it even more difficult to remove.

For washable clothes, the best way to get rid of grease marks (from mayonnaise to machine oil) is to wash them with detergent in the hottest water possible. Pre-wash treatment with a proprietary stain remover will also help, but be sure to test the fabric first to make sure it is colourfast. Take anything that can't be washed to the dry cleaners and highlight the offending areas for special treatment. For floors, carpet shampoos are most effective on grease, either for spot or all-over cleaning.

Candle wax can be more easily scraped off if it is first hardened by chilling with a block of ice. What remains can then be given the hot iron and blotting paper treatment.

On grease, detergents are much more effective than soap because they not only act to emulsify the fat – breaking it up into minute droplets, which will then lift off the fibres of a fabric – but, as they do so, they attach themselves selectively to dirt. The first detergents were developed in Germany in the 1880s, and Nekal was the first branded detergent, sold there in 1917.

In the days before detergents, another recommended method was to rub dry flour on a garment or carpet and to leave it for several hours to absorb any grease or oil. Turpentine was also used as a dry cleaner.

WARM WINE WILL MAKE JEWELLERY SPARKLE

Certainly a means of cleaning silver jewellery, but best done with white wine, which will not stain. Warm water mixed with soap or a little mild washing-up liquid is, however, much safer for most types of jewellery.

It is wise to choose your jewellery cleaning method carefully, but in every instance a soft toothbrush or eyebrow brush is useful for getting into the crevices. For gold, diamonds and other hard stones such as emeralds, rubies and sapphires, there is no need for anything more than warm soapy water, but soft stones such as amber and opal should simply be wiped with a silk cloth. Silver is best cleaned with a soft cloth dipped in bicarbonate of soda or a commercial paste or liquid polish, then rinsed in warm soapy water and dried.

Pearls need careful treatment – ideally nothing more than a gentle rub with a soft chamois cloth after they have been worn. Dipping them in a lukewarm, very dilute soap solution, then quickly rinsing and drying them, will help to restore their lustre, but their strings should never be allowed to get wet. A cameo, after cleaning in soapy water, will appreciate treatment with a little olive oil or oil of wintergreen, to prevent it from becoming dull and cracked.

The earliest jewellery, worn long before clothes became customary, was made from wood, bones or the tusks of creatures such as mammoths. In Europe, the oldest ornaments known date to around 38,000 BC. Gold was first discovered some time before 3000 BC, and some of the oldest known gold jewellery was crafted in Sumeria.

Never rub soap into woollen garments

A wise precaution when dealing with wool, which needs careful treatment to preserve its texture and to prevent it from shrinking. Keeping woollens scrupulously clean is an excellent deterrent to moths.

The House and Home Practical Book of 1896 is specific in its instructions for dealing with wool: apart from the use of ammonia, the advice holds good today. 'Dissolve the soap in water. Do not rub woollens on a board. That process mats the wool. Rub in the hands. Rinse in water in which there is ammonia of a temperature that might be called a little more than warm. Rinse until the last water is perfectly clear. Dry in a temperature as near that of the water used as possible.'

For coloured woollens, *The Practical Housewife* prescribes a method that seems more suited to the kitchen than the laundry, although it is appended with a personal recommendation from a Mrs J.D.R: 'Quarter pound of soft soap, a quarter pound of honey, the white of an egg, and a wine-glassful of gin; mix well together, and the article to be scoured with a rather hard brush thoroughly, afterwards rinse it in cold water, leave to drain and iron whilst damp.'

Wool is one of the oldest yarns known. Spinning the fleece of sheep into wool probably began in around 9000 BC in Mesopotamia, where the animals were originally domesticated. Until the invention of the spinning machine some 2,000 years later, the yarn would have been laboriously rolled by hand.

Advances in fibre technology have resulted in a 'deep immersion' process in which wool fibres are soaked in a shrink-proofing solution. This results in machine-washable wool that holds its shape and integrity while retaining wool's unique qualities of warmth, softness, breathability and natural comfort.

A WHITE DOORSTEP IS A CLEAN DOORSTEP

A saying that goes back to the 19th century, when it was customary – and a matter of pride – to whiten the front doorstep as a mark of cleanliness. In northern England the practice continued right up to the 1960s.

> A doorstep is a colloquial name for a sandwich made with thick slices of bread.

The doorstep was whitened each day, after it had been scrubbed, using a bleaching agent called stone blue mixed with size, whiting and pipeclay. Once dry it would be rubbed down with a flannel and then brushed. The task had to be performed daily as the whiting wore off easily when it was walked on.

In a Victorian house with servants, cleaning the doorstep would have been one of the duties of the scullery maid or the maid-of-all-work. Describing the daily duties of the latter – after opening shutters, lighting fires, laying breakfast, cleaning the grate and dusting, Mrs Beeton says: 'The hall must now be swept, the mats shaked, the door-step cleaned, and any brass knockers or handles polished up with the leather.'

In Greece and other Mediterranean countries, doorsteps were whitened with a mixture of lime and asbestos, which was potent enough to kill insects such as ants. As well as doorsteps, the mixture was painted on to courtyards and walkways, and even on tree trunks.

TACKLE A DIRTY OVEN WITH BICARBONATE OF SODA AND VINEGAR

A reasonable way of getting rid of grease in an ordinary oven that is not self-cleaning, but far from totally effective on stubborn, baked-on deposits. Tedious though it may be, wiping out the interior of the oven with a dilute bicarbonate solution every time it is used is the best way of keeping it in tip-top condition.

Oven cleaning is a modern chore linked to the advent of gas and electric cookers. Originally the interiors of these ovens were coated with enamel that needed extremely careful treatment to prevent abrasion, and even its modern equivalents respond best to a gentle touch. For this, mix a good quantity of bicarbonate of soda (baking soda) with enough water to make a paste. Lightly moisten the oven interior then apply a thick layer of the paste and leave

it for at least three hours. Next, spray the entire oven with white vinegar using an ordinary spray bottle. When this hits the bicarb it will fizz and help dissolve the grease, though it will need to be left for a few more hours before being wiped off with a sponge dampened with hot water.

It is better to prevent the oven becoming too dirty in the first place, and *The Gas Cooker Cookery Book* of the 1930s gives a handy summary of oven care: 'After roasting, the oven should be washed down with a cloth wrung out in soapy water before the enamelled linings get cold. It is much easier to clean the oven

'The best and most expeditious way to remove grease from a cooking stove is to rub the mark as hard as possible with a good supply of ordinary newspaper. If this is done when the stove is still warm the grease stains will come away quite easily.'
(Mrs E.W. Kirk, 1924)

immediately after use whilst the enamelled plates are still warm. Grease splashes become baked on the enamel if the oven is used again before cleaning off and are more difficult to remove.'

Caustic commercial oven cleaners are usually very strong and need to be applied with great care. If you have to use one, wear rubber gloves, spread newspaper on the floor in front of the oven and be sure to prevent the mixture from touching your skin or other kitchen items.

CHARCOAL WILL ABSORB ANY EFFLUVIA

A tribute to charcoal's ability to absorb noxious gases and therefore act as a deodorant. In wartime, activated charcoal, treated so as to greatly increase its surface area, was a key constituent of gas masks.

Charcoal is the porous black, brittle substance that remains when wood or bones are charred or partially burnt. Historically, it has been most widely used as a fuel, prized for the fact that it burns more brightly and hotly than wood, with little smoke – facts still appreciated by those who cook on a charcoal barbecue.

A typical use of charcoal in the Victorian home was to put pieces of it in a wardrobe or chest of drawers to keep linen smelling fresh. It was also used for filtering putrid water, 'the object being to deprive the water of numerous organic impurities diffused through it, which exert injurious effects …' Though our tap water today is pure enough to drink, charcoal-based filters are extremely popular for removing the chlorine added to disinfect it as well as excesses of minerals such as calcium carbonate, which make the water very hard.

When water closets were first introduced to the home they were decidedly smelly, and charcoal ventilators – consisting of a thin layer of charcoal set between sheets of wire gauze – were used to purify the air. They were also set over sinks and drains.

> An old remedy for preventing a foul odour when a pot or pan boils over is to throw some salt on to the stove.

Face masks, constructed in a similar way, were used in hospitals and sickrooms as a means of reducing the risk of infection. The first proper gas mask to be employed in warfare was invented by the American chemist James Bert Garner in 1915; it was subsequently used on the Western Front.

CHARCOAL'S PURIFYING POWERS

For health, charcoal was once employed widely and in many guises. *Enquire Within* includes the following references:

Powdered charcoal applied to sores will 'arrest the progress of gangrene'.

'Meat, poultry, game or fish, &c., may be preserved for a longer period in hot weather by sprinkling it with powdered charcoal.'

'A little charcoal mixed with clear water thrown into a sink will disinfect and deodorize it.'

'Rubbing the teeth and washing out the mouth with fine charcoal powder will render the teeth beautifully white and the breath perfectly sweet …'

THE KNITTED DISHCLOTH IS EXCELLENT FOR USE AND WEAR

The dishcloth is a hard-wearing, old-fashioned item that still earns its place in the kitchen. For good hygiene, dishcloths must be kept scrupulously clean.

Thrifty tip: for cleaning into small crevices that the dishcloth will not reach, save and use old toothbrushes.

The magazine *Home Chat* for 16 May 1896 informed its lady readers that knitted dishcloths are easily made: 'Four skeins of knitting cotton will be required. Cast on thirty-six stitches, and knit a square with large wooden needles.'

In use, the dishcloth is needed to remove food remains from cutlery and plates. To protect your silver and china it is best to use a firm rotary motion and to resist the urge to scrub, which can cause damage especially if dishes are decorated with gilt. Adding a little white vinegar or bicarbonate of soda to the washing-up water can, if necessary, help to dissipate excess grease.

The Victorian kitchen maid would have kept her dishcloths clean by rinsing them, then boiling them in a pan of water. The 21st-century kitchen offers a much easier alternative. Zapping them in the microwave on high for two or three minutes will kill 99.99 per cent of any bacteria that may be lingering. Make sure, however, that the cloth is very wet – if it is dry you risk starting a fire.

HARD WATER WILL NEVER MAKE A LATHER

Soap certainly lathers best in soft water. Hard water, which contains lime and other impurities, prevents soap – and detergent – from making a good foam. It also deposits limescale on sinks, kettles and all kinds of other domestic appliances.

Apart from being more economical with soap and detergent, soft water is also much more pleasant on the skin. Best of all is rainwater, which is soft because, unlike water from earth-bound sources, it has not had the chance to absorb minerals such as calcium and magnesium from contact with rocks.

> It is said that for a man 'a good lather is half the shave', but to get into a lather is to lose your cool.

Washerwomen of old would soften their water with salt or by adding borax (see page 110) or soda. Salt – sodium chloride – is still used today in water softeners, which work by flushing salt through a chemical matrix called zeolite. As hard water flows through the matrix, calcium and magnesium ions are exchanged for the sodium ones and the water is softened.

The water that comes out of our taps, although it has almost certainly been recycled, is pure and drinkable. This was far from the case in Victorian times when, as *Household Science* of 1889 pointed out, 'there are other impurities in water more dangerous than the inorganic ones. These proceed from decaying animal and vegetable matter (as from sewage), which has found its way into it.' For protection, the book recommended that there should be 'a filter in every house if possible, and all water used for food purposes whatever should be filtered.'

To get your washing white, dry it in the sun

Apart from the lightening effect of sunlight, oxygen and ozone in the air work as bleaches, which not only whiten but disinfect the wash – not to mention the added benefit of the wonderfully fresh aroma of sun-dried linens. On the down side, washing left outside to dry is at the mercy of the weather.

If there is a light breeze, hanging the washing outdoors can also help remove creases from clothes and cut down on the ironing. However, thick items such as towels can get hard and stiff when sun dried in still air, and may be better finished off in a tumble dryer or airing cupboard.

There is an art to hanging out the washing. First you need to choose what to pin out: woollens, silks and delicate fabrics are better kept indoors and dried flat. It would have been second nature to the laundry maid to peg socks by the toes, shirts by the tails and dresses by the shoulders. Adjacent tea towels and handkerchiefs she would have pegged together at the corners, making a neat, firm row on the line.

In seafaring communities, wives will never wash clothes on the day their menfolk set sail, for fear that doing so will wash their ships away.

For securing items such as sheets and jeans so that they don't blow away there is still nothing to beat the traditional push-on or American peg made from a single piece of wood. This is not only durable but, unlike the spring-grip peg (which comes in plastic as well as wood and is most suitable for light, more delicate items), has no metal to rust. The gypsy peg, rarely seen nowadays, is made from two pieces of whittled wood held together at one end with a band of metal.

In patriotic defiance at the outbreak of war in 1939, songs such as 'We're Going to Hang Out the Washing on the Siegfried Line' became popular in Britain.

THE INSIDE OF A BANANA SKIN WILL BUFF UP BROWN LEATHER

An old – and a quick and economical – way of getting a shine on both black and brown leather, but not to be recommended for paler colours. Before proprietary shoe polishes were widely available, mixtures would often be made at home.

> The cunningly placed banana skin – the prop of the slapstick comedian – is the source of slips and falls.

Now, as then, all shoes and boots benefit from being kept clean. Any loose dirt or mud needs to be removed with a stiff brush before shoes are polished. Leaving the polish to dry for about 10 minutes, rather than rubbing it off straight away, will help to preserve the leather and add extra shine. White shoes are best cleaned with a cloth dipped in a little ammonia. For patent leather you can use a proprietary cleaner or simply remove any marks with an eraser then buff up the leather with a little baby oil or petroleum jelly.

Ivory black – made from charred animal bones – was an essential ingredient for home-made shoe polish. *Enquire Within* gives this potent-sounding recipe as 'Best Blacking' for boots and shoes: 'Ivory black, one ounce and a half; treacle one ounce and a half; sperm oil, three drachms; strong oil of vitriol, three drachms; common vinegar, half a pint. Mix the ivory black, treacle and vinegar together, then mix the sperm oil and oil of vitriol separately, and add them to the other mixture.'

It was an old practice to rub new boots with the cut side of half a lemon, then leave them to dry thoroughly in order to produce a good surface for polishing. For cleaning leather chairs, a mixture of one part vinegar mixed with two parts of boiled linseed oil was recommended.

CLEAN KID GLOVES WITH BREADCRUMBS

One of the many household hints presented to young women in *The Girl's Own Paper* of the 1890s, which also pointed out, quite correctly, that such gloves get very dirty inside long before they are worn out.

To clean kid gloves by this method they first need to be turned inside out. Then take a piece of stale bread and rub it firmly over the surface. The crumbs that become detached as a result will take with them the dirt from the leather. Whichever cleaning method you choose, wearing gloves on the hands one at a time is the easiest way of dealing with them. While kid, a light and delicate leather made from goat or sheepskin, will shrink if it gets soaked with water, it is safe to sponge it with a damp cloth and to let it dry slowly in the air. Any marks on the outside of the leather may respond to treatment with a clean pencil eraser. A sprinkling of talcum powder can also help to absorb stains caused by grease or oil. For suede, a very fine wire brush or fine sandpaper may be used, but check first on an unobtrusive area that it will not do any damage.

Anything that fits 'like a kid glove' is perfect for size, and to be 'hand in glove' with someone is to be as close as possible, while a sensitive and temperamental person may need to 'handled with kid gloves'.

Splits in kid gloves are hard to mend, since simply stitching the edges of the leather together will tighten them and make them split again. However if you work a buttonhole stitch around the split, using a thread in the same colour as the leather, then buttonhole stitch again, using the loops of the first round of stitching as your anchors, the edges will hold firmly together.

To keep kid or leather gloves in good shape, avoid pulling them violently at the wrists when you put them on. Equally, when taking them off, pull gently at each fingertip and the thumb, using as little force as possible

PUT MILK ON AN INK STAIN

One of many remedies favoured in past times, but by far the least effective as it may in fact 'set' the stain permanently. Methylated spirits – or water for washable fountain pen ink – are much better alternatives.

Washable ink is best removed with water and a mild detergent. For ballpoint ink stains on both clothes and carpets, experts recommend methylated spirits applied on a cotton bud or cloth (but only after a piece of fabric has been tested for its durability and colourfastness). After the meths treatment, rinsing in a 1:10 solution of white vinegar, then plain water, should remove the stain completely.

The oldest known inks – used for creating pictographs or 'word pictures' on cave walls – were made from charcoal for black and iron oxide for red, mixed with animal fat. The word 'ink' comes from the Greek *enkaiein*, 'to burn in', and refers to the practice of heating mixtures to fix them.

Early inks were also made from plants. In one 12th-century recipe, hawthorn branches were cut in the spring and left to dry. The bark was then removed, pounded and soaked in water for eight days, after which it was boiled with wine until the mixture thickened and turned black. This was poured into bags and hung up to dry before the resulting powder was heated with more wine and iron salts.

Other unlikely ink stain removers include tinned tomato juice and tallow.

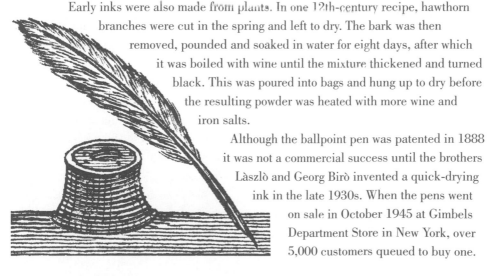

Although the ballpoint pen was patented in 1888 it was not a commercial success until the brothers Làszlò and Georg Birò invented a quick-drying ink in the late 1930s. When the pens went on sale in October 1945 at Gimbels Department Store in New York, over 5,000 customers queued to buy one.

SALT MAKES AN EXCELLENT DRY CLEANER

This once popular method has been superseded by commercial dry cleaning, but it may still be worth trying as long you have checked that it will not do any damage to the garment. The great value placed on salt over the ages is reflected in many superstitions.

For dry cleaning at home Mrs E.W. Kirk, in her *Tried Favourites* of 1924, prescribes this method: 'Lay one penny lump of salt in oven till warm. Well shake and brush the garment and lay on table. Break off piece of salt, and rub on article till salt is soiled. Continue till it is clean, then shake well and brush, and place in air or before fire to take out creases.' It was recommended that, if necessary, the application be repeated two or three times.

Other, much more dangerous dry cleaning procedures were used in the same era. For silk and satin, petrol was applied and the fabric rubbed with a folded pad of white cloth, then squeezed to remove any remaining petrol. However *The Concise Household Encyclopedia*, in its description of this method, does include the warning that 'on no account' should it be used 'except in the open air'.

If you spill grease or oil on your worktop, scatter salt over it at once and it will be easy to clean.

The oldest means of dry cleaning, used by the Myceaneans around 1600 BC, was to sprinkle dirt on clothes, then brush it off. Even more effective was fuller's earth, a kind of clay. Thanks to its ability to absorb grease and dirt it has been used for cleaning wool for over 7,000 years. In the traditional process of cloth finishing, lanolin and other oils and dirt were removed by kneading the cloth with fuller's earth and water, and it was then stretched and pounded to make it thicker and stronger. The whole operation was known as 'fulling', hence the name of the clay. Even in the 1930s fuller's earth was still being used for dry cleaning rugs and carpets.

For spills in the oven, adding some water to dampen them, plus a liberal sprinkling of salt, will make them much easier to scrape away when dry. Salt will also refresh the inside of a coffee pot or teapot.

SALTY SUPERSTITIONS

Reasons why you should treat salt with respect:

Salt scattered on to a field will ensure a good harvest.

If your home is short of salt you will also be short of money.

Carrying salt in your pocket is a protection against evil – and especially witchcraft.

When you move into a new home, sprinkle salt into every room to ensure good luck.

At table, do not help your neighbour to salt. This is tantamount to helping them to ill fortune.

If you spill salt, throw a handful over your left shoulder and you will be spared ill luck.

Never lend your salt, or ask for it, for to part with salt is to part with luck and to ask for salt is to ask for sorrow.

PUT WHITE WINE ON A RED WINE STAIN

Red wine on a pale carpet or favourite sweater is the accident everyone dreads, but white wine is better drunk than wasted on a stain! Salt may work to a degree (and is good for bloodstains) but a mild detergent is by far the best treatment.

When poured on to a stain, the bubbles in soda water can help lift a red wine stain from carpet or fabric fibres, but whatever remedy you choose, haste and a gentle touch are essential. To remove wine stains from linen, a common 19th-century suggestion was that you should hold it in milk 'while boiling on the fire'.

Your main aim should be to prevent the stain from 'setting', spreading or being rubbed into the fibres of a carpet or a garment; to this end it is important to mop up as much as possible of a spill. The advantage of the detergent approach is that you can whizz up a foam, which can then be palmed on to a carpet to work on the stain without getting the pile soaking wet. The residue can then be carefully removed with a clean cloth, without rubbing, and the area neutralized with a very dilute solution of white (not wine or malt) vinegar. The same treatment can be tried with spilt coffee or tea. If all else fails, professional cleaning may be your only hope.

Detergent also helps with other potentially difficult substances. With protein-based stains such as blood, egg, milk or chocolate, dilute the detergent in cold water to avoid 'cooking' the stain.

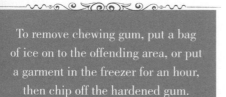

To remove chewing gum, put a bag of ice on to the offending area, or put a garment in the freezer for an hour, then chip off the hardened gum.

FOR A GRASS STAIN, RUB ON PLENTY OF TREACLE

… and afterwards wash the garment in tepid water. So says Mrs E. W. Kirk in her *Tried Favourites*, but it is not a method that would be recommended today. Grass stains remain among the hardest to eliminate.

Grass stains are so stubborn because the grass contains the green pigment chlorophyll, which, along with other plant juices, binds to the fibres in clothes. What is important is to avoid setting the stain by trying to clean it with substances such as ammonia or alkaline detergents, which will make it virtually impossible to remove. Ideally, treat the stained area with a commercial stain remover, then rub some liquid laundry detergent into the garment before soaking it in dilute bleach.

A good home-made pre-wash treatment can be made by mixing equal parts of white vinegar and warm water – the acid helps to dissolve away the stain. Housewives of previous generations would also use Javelle water, purchased from

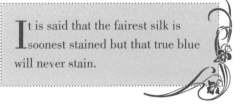

It is said that the fairest silk is soonest stained but that true blue will never stain.

the chemist or made by shaking together in a bottle 30oz water and 2oz chlorinated lime or bleaching powder, then adding 4oz carbonate of potash in 10oz water. This mixture was left to stand for several days then strained to get rid of any sediment. Because of its bleaching properties, Javelle water was suitable for use on white garments only. It would also quickly rot fabrics unless rinsed off very thoroughly.

STAINS ON MARBLE ARE NOT EASILY REMOVED

They also need to be removed very carefully, because marble, although it appears hard, is in fact a delicate and porous substance. Marble is one of the most beautiful materials, used for millennia in the decorative construction of homes.

In the game of marbles, clay, stone or glass, rather than marble itself, are used to make the balls used in playing the game.

Soap and water – as for the skin – is a perfect cleanser for marble. The soap should be the gentlest you can find and dissolved in about 2 litres (½ gallon) of warm water. The marble can then be carefully sponged or mopped. *Our Homes* of 1883 suggests that if the marble is 'very dirty, a stiff paste may be made with soft soap, caustic potash lye and whiting. This,' it advises, 'should be spread upon the marble and allowed to remain a while, then taken off, when the marble can be washed with soap and water.' Today, a thick paste of bicarbonate of soda (baking soda) mixed with water is advised as a gentler alternative.

To keep a marble floor in good condition it is wise to polish it regularly with a polishing powder such as tin oxide or a damp cloth dipped in powdered chalk. The floor should also be sealed annually with a proprietary stone sealer. *Cassell's Household Guide* gives a recipe for a polish for marble: 'Melt over a slow fire four ounces of white wax, and while it is warm stir into it with a wooden spatula an equal quantity of oil of turpentine; when thoroughly incorporated, put the mixture into a bottle or other vessel, which must be well corked whenever not in use. A little of the above is put upon a piece of flannel and well rubbed upon the marble.

Marble is a type of limestone, which lends itself to the creation of beautiful surfaces, from floor to pillars, as well as for sculpture. Magnificent examples of Roman marble floors still exist in the excavated remains of Herculaneum, the town buried by lava from the eruption of Vesuvius in AD 79. The purest marble is said to come from Carrara in Italy – the source of the stone used by Michelangelo for his *David* – and from Villa Vicosa in Portugal. The best of British marbles include mottled grey Purbeck and the black and red Derbyshire marbles. A fine blue-green marble is quarried in Connemara in Ireland.

Any chips in white marble can be filled with plaster of Paris, although you should bear in mind that it is virtually impossible to match for colour.

CLEAN COPPER WHILE IT IS STILL HOT

Not absolutely necessary, although it is always good to clean pots and pans as soon as they are finished with. While they demand much care, copper utensils are invaluable in the kitchen, notably for whisking egg whites.

The way that copper is best cleaned depends on whether or not it has been lacquered. If lacquered, as is the case for most decorative household items such as vases and figurines, it can be buffed up with a duster or wiped over with a damp cloth. Cleaning copper cooking utensils is vital because if left dirty and damp poisonous salts, including verdigris and copper carbonate, can form on the metal.

Unlacquered copper, which is the norm for cookware, should never be cleaned with a scouring pad or anything that will scratch it. Equally, any cleaning fluid containing bleach risks discolouring the metal permanently.

For ordinary cleaning, sprinkle a copper pan with salt and some white vinegar, or cut a lemon in half, dip it in salt, and rub it over the surface. This is also a good treatment for so-called 'bronze disease', which creates patches of corrosion. If the pan is tarnished, simmer it for several hours in a large pan of water containing a teaspoon of salt and a cup of white vinegar. Alternatively, make a paste of lemon juice and cream of tartar, spread it over the copper surface, leave it for 5 minutes then wash in warm water and dry.

In the kitchen of times past the 'copper' was a large boiler, used for both laundry and cooking. It was so named because it was originally made of copper, though this was eventually superseded by galvanized iron. It was heated from below by a gas burner.

Copper utensils look lovely in the kitchen and are wonderful conductors of heat. However, even if lined with tin, they should never be used for cooking acidic foods.

For beating egg whites, copper is best because – as cooks knew of old and researchers have now proved in the laboratory – there is a reaction between the copper and conalbumin (one of the proteins in the white) that prevents the foam from separating out into a nasty mess of lumps and liquid and imparts a creamy yellow colour quite different from the snowy white of a foam whisked in a glass, stainless steel or ceramic bowl. A copper bowl produces a stable foam that can be whisked stiff and will stay firm and not collapse, but whatever it is made of, a clean, dry container is essential to a good result.

SCRUB THE BATH BEFORE YOU LEAVE IT

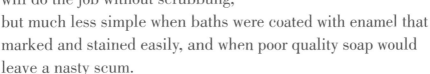

Easy enough today, when just a quick spray with a bath cleaner will do the job without scrubbung, but much less simple when baths were coated with enamel that marked and stained easily, and when poor quality soap would leave a nasty scum.

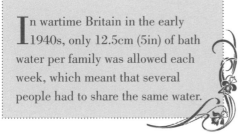

In wartime Britain in the early 1940s, only 12.5cm (5in) of bath water per family was allowed each week, which meant that several people had to share the same water.

To remove unsightly marks and grubby rings caused by soap scum on an enamel bath, Mrs E.W. Kirk, writing in 1924, said that the best treatment was to 'dip a flannel moistened in paraffin oil into salt and rub vigorously on the bath; then rinse with plenty of hot water and dry. If the marks are not of long standing,' she adds, 'they can be removed with salt alone, or with salt moistened with spirits of turpentine. Apply this to the bath when dry.' Today, detergent applied undiluted and left overnight will probably do the trick, while for day-to-day cleaning a proprietary enamel-safe cleaner will do the job well.

Baths can suffer from stains. Mild rust stains should disappear with lemon juice applied on an old toothbrush, while a paste made from equal amounts of cream of tartar and bicarbonate of soda mixed with lemon juice and spread on to blue-green stains should be effective if left for half an hour then rinsed off with warm water.

Commenting on the positioning of the bath in the 1930s, before the advent of the fitted bathroom, one household guide says: 'It should be quite clear of the walls at the sides and ends, so that there is no difficulty in being able to clean and dust all round it. It should also be raised from the floor with wooden supports, if the legs are not long enough to permit of sweeping under it.'

POLISH FURNITURE WITH BEESWAX

Beeswax is the most natural and still one of the best ingredients for keeping wooden furniture in great condition, though special finishes need other kinds of treatment.

> Beeswax is produced by worker bees, which secrete it from their bodies. It is created from a mixture of honey and pollen.

Waxing and polishing furniture protects it and keeps it looking shiny and bright. And once waxed, it won't need re-treating more than once or twice a year, depending on how often it is used. Once well waxed, with a bought or home-made polish (see below) it can be regularly cleaned with a duster (aided if you wish by a proprietary dusting spray) or wiped with a barely damp cloth. Furniture that is finished with shellac, lacquer or varnish is better left unwaxed. Soapy water used sparingly on a soft cloth should remove all but the worst marks.

You can make a good all-purpose beeswax mixture at home by melting 2 tablespoons of beeswax granules with the same quantity of turpentine. Oak furniture that is stained or dull will take on a wonderful shine if polished with a warm mixture of 1 tablespoon beeswax granules melted with 300ml (½ pint) beer and 2 teaspoons sugar.

Camouflage is the best way of dealing with scratches and other marks on polished wooden furniture. Buy one of the 'touch up' pens sold for the purpose or try a small amount of artist's oil paint in a matching colour, applied with a cotton bud or a fine paintbrush, or some cream shoe polish applied in the same way.

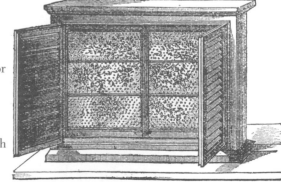

A NAILBRUSH WILL CLEAN A COMB MOST EASILY

An old toothbrush is also handy for the same job, which needs to be done regularly using a mild detergent or soapy water. Decorative hair combs, made of bone and dating back to at least 8000 BC, are among the most ancient accessories known to human society.

Women of ancient Egypt used combs made from ivory and wood with teeth on both sides. Plastic combs have been made since 1862, when the chemist Alexander Parkes created the plastic Parkesine by mixing chloroform and castor oil, but while modern plastics are excellent, tortoiseshell remains the finest material for a comb.

If you drop a comb it is believed to be unlucky to pick it up yourself unless you tread on it first.

On choice, *The Concise Household Encyclopedia* says that: 'Dressing combs should not be too cheap, or the teeth will break readily: also roughness in finish irritates the scalp. The teeth should be of two different sizes, coarse at one end and fine at the other, and the comb should be discarded as soon as it is damaged.' Modern comb cleaners, consisting of small rotating brushes, can be purchased and, like the nailbrush, are much easier to use than the 1930s version consisting of a series of threads on a small frame. Also made in this period were bizarre battery-operated hair combs, which created a circuit when they came into contact with the scalp.

New brooms sweep clean

Another way of saying that change is a good thing, and usually refers to the way a new administration clears away old routines and brings in new ideas. The housewife can also take it literally as a reminder to renew her cleaning utensils regularly.

Appropriately, the broom shares its name with the flexible branches of the plants long used for cleaning (botanically the genera *Cytisus* and *Genista*), for it was these, as well as branches of birch, heather and tufts of maize, that were probably employed to sweep the floors of early dwellings. Tied together, and with a handle added, bundles of these twigs became the besom. Though good for dispersing worm casts on lawns and sweeping up leaves, the besom is relatively inefficient for household cleaning. In the USA, a flat, stiff broom with long bristles made from corn husks was invented by the Shakers in the 19th century, and helped to collect dust and dirt rather than scattering it about.

The wooden push-broom, with bundles of bristles glued into its head and a broomstick attached at an angle, dates back to the 15th century, and was still vital to the cleaning repertoire long after the invention of the vacuum cleaner. Brush purchasers of the 1930s were advised to look out for the presence of 'inferior mixtures' hidden in a broom, which would adversely affect its cleaning powers. These, it was advised, should be identified 'by placing the hand across the surface of the broom and noting the greater readiness with which the bristle will spring back when released, as compared with the less springy substitute.'

Good brooms deserve to be well looked after. The efficient housemaid would have turned the heads of her brooms regularly to make sure that they wore evenly. She would also put them into boiling water once a week to make them tough and durable.

SWEPT AWAY

Dozens of old sayings relate to the buying and handling of brooms – apart, of course, from their use as transportation for witches and their feline familiars:

Brooms bought in May sweep the family away.

If you set a broom in the corner, strangers will come to the house.

Lay a broom across the doorway to protect the house.

Throw a broom to ward away witches.

It is an ill omen to put a sweeping brush on a table.

A servant will not get her wages if the head comes off her broom while she is sweeping.

CHAPTER 5

A WELCOMING HOME

The perfect home is so much more than a place for eating, sleeping and watching television. It is a place where friends and neighbours, as well as members of the extended family, should always feel welcome. Whether entertaining guests for a meal or inviting them to stay overnight – or longer – it is the role of the host and hostess to make sure all is in order, and every detail is taken care of.

Entertaining guests for meals today is likely to be much more informal than it would have been even 50 years ago, but it is still worth knowing how to set the table, arrange candles and floral decorations, work out a seating plan, and serve food and wines so that they are enjoyed to the full. Even when it is shared around the kitchen table, good food still deserves to be properly served, whether it is cheese brought to room temperature or a soup prepared with care from fresh ingredients. During dinner good guests will know how to make stimulating but not contentious conversation. After the meal, host and hostess will continue to ensure that their guests are amused for the remainder of the evening.

For overnight guests – once said, like fish, 'to smell in three days' – even more attention is required. Beds must be made and well aired and guest bedrooms supplied with everything from freshly laundered towels to a selection of reading matter. For outdoor activities the perfect hostess will still keep a supply of umbrellas for her guests and, in the country, add a selection of Wellington boots and walking sticks.

HOSPITALITY IS A MOST EXCELLENT VIRTUE

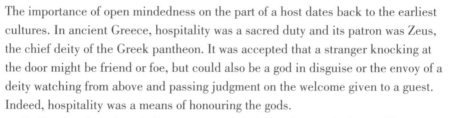

But also makes demands on hosts to ensure that guests feel welcome. The ceremonies of hospitality almost always involve sharing food and drink.

The importance of open mindedness on the part of a host dates back to the earliest cultures. In ancient Greece, hospitality was a sacred duty and its patron was Zeus, the chief deity of the Greek pantheon. It was accepted that a stranger knocking at the door might be friend or foe, but could also be a god in disguise or the envoy of a deity watching from above and passing judgment on the welcome given to a guest. Indeed, hospitality was a means of honouring the gods.

In Greek culture hospitality was centred, as it is today, on the home. The master of a household formed allegiances with those of other households and, as result, they all grew in wealth, strength and stature. Violations of the unwritten rules of hospitality could have dire effects, not least the inexorable wrath of the gods: it was believed that the earthquake of 373 BC, which destroyed the city of Helike, was sent to punish its citizens after they refused hospitality to the Peloponnesians.

> The word hospitality comes from the Latin *hospes*, which originally meant 'stranger' and has the same root as *hostis*, a 'hostile stranger'. The implication is that the true host will accept all comers.

In the Bible, the notion of extending hospitality to all is expressed by Job, who says, 'No stranger ever had to sleep outside, my door was always open to the traveller.' It is also epitomized by Abraham, who showed unreserved hospitality to guests, and by the Good Samaritan, who, in Luke's Gospel, having bandaged the wounds of the man set upon by robbers on the road from Jerusalem to Jericho, 'lifted him on to his own beast, brought him to an inn, and looked after him'.

THE FRONT DOOR IS A HOME'S MOST IMPORTANT FEATURE

The front door of a house needs to be substantial, both for effect and for security. As the entrance to the home, it should be welcoming, but it should also provide protection against undesirable visitors.

In times past, well-to-do households would have had servants to answer the knock at the front door or the ring of the bell. Pondering the difficulty that lack of servants presented, Mrs Caddy in her 1877 book *Household Organization* asks, 'Who is to answer the door?' and 'Who will be the slave of the ring?' Analysing the subject, she identifies 'four classes of people who knock at our door: the family, tradespeople, visitors, and casuals' and lets her mind roam to the possibility of a 'turnstile door, which would allow the easy delivery of goods by tradesmen'. As to

the issue of visitors she concludes that the best remedy is 'to have no friends but those whom we are glad to see ...'

The electric doorbell was invented in the early 20th century, and by the 1930s musical chimes were all the rage, as this 1937 American advertisement from Rittenhouse testifies: 'When the door-button is pressed, two rich, clear chime tones replace the irritating, nerve racking noise of the ordinary bell or buzzer. An artistic addition to any room. AN IDEAL CHRISTMAS GIFT.'

Before homes were well heated, it was customary to hang a thick curtain in the hallway to cover the front door.

In *From Kitchen to Garret* Mrs J.E. Panton recommends 'a double curtain of serge or felt. This, the author says 'could be arranged on one of those delightful rods that are, I believe, only to be purchased at Maples [a big London furniture store]. And that move with the door itself in some mysterious way …'

Tradition Associates the Open Fire with an Open Heart

There are few features more welcoming in the home than an open fire, and the fireplace remains the best focus for any sitting room. It is said, however, that if you enter a neighbour's house and the mistress is poking the fire, you should take it as a sign that you are not welcome.

According to the best archaeological evidence, the caves inhabited by Peking Man some 35,000 years ago were lit by fires. The tradition of having a central fire in the main room of the home, surrounded by stones and with an outlet for smoke in the roof, persisted for many millennia. In Britain it was only in the 14th century, after chimneys were integrated into the walls of buildings, that the fireplace moved from the centre to the side of the room.

Wood was the first fuel for domestic fires. As coal came to replace it in the 18th century, metal fire grates were developed to burn it. With the addition of flat supports on which to place pots and pans, food could be cooked more easily on an open fire, and from such beginnings the kitchen range developed.

When tending a fire, the 'use of the poker,' says *Enquire Within*, 'should be confined to two particular points – the opening of a dying fire, so as to admit the free passage of air into it, and sometimes but not always, through it; or else, drawing together the remains of a half-burned fire, so as to concentrate the heat, whilst the parts still ignited are opened to the atmosphere.'

FIRED FOR FORTUNE

Many rituals and superstitions relate to the treatment of the fire in the fireplace:

You must be a friend of seven years' standing before you take the liberty of poking the fire in someone else's house.

Never let the fire go out on New Year's Eve, Halloween, Midsummer Eve or Christmas Eve: you will give away your good luck.

If two women try to light a fire together they are sure to quarrel.

A fire that spits and sparks is a warning of a reprimand or scolding to come.

A fire that is still alight in the morning foretells an illness or even death.

Blue flames in a fire betray the presence of the spirits of the hearth.

SILENT PLUMBING IS THE MARK OF A WELL RUN HOME

An ideal that is virtually impossible to achieve, but worth aiming for. The noise of flushing toilets is inevitable, but no one will want to endure the sound of banging pipes at night.

A phenomenon called water hammer is a common cause of banging pipes. This happens when fast-moving water in the pipes is brought to an abrupt halt. In well-installed plumbing, air chambers absorb the shocks and prevent the hammering, but if these fail then attention is needed. If possible, you need to turn off the water behind the waterlogged chamber, open the tap and allow the pipe to drain thoroughly. If this is impossible, or fails to have any effect, then you will need to call in a plumber.

Take good care. 'Water closets need very careful attention. They get out of order mostly by having improper substances thrown down by idle or careless persons.' (*Household Science*, 1889)

Noisy plumbing also occurs because of poorly fitting ball valves in toilets and because pipes are so loosely fitted that they bang against the walls as water flows through them. Pipes that bang against brick walls can be stabilized and silenced by fitting a block of wood between pipe and wall and fixing it with a metal bracket.

The flushing toilet was invented in 1499 by Thomas Brightfield of London, in an age when chamberpots were emptied daily into the streets. In 1596 Sir John Harington installed a flushing water closet in his home near Bath, but such conveniences became universally available only with the advent of public sewerage systems. The water closet with an S-bend designed to prevent noxious odours rising was invented by Alexander Cummings, a watchmaker, in 1775, but the real advance came when the plumber Thomas Crapper invented the valveless water cistern in 1884.

THE HOME IS NOT A BATTLEFIELD

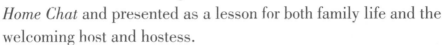

Advice given to women under the heading 'things worth knowing' in the Victorian women's magazine *Home Chat* and presented as a lesson for both family life and the welcoming host and hostess.

Compared to today, running a large house complete with servants was an arduous task for any woman – who was also expected to be able to greet her husband and any guests with magnanimity. Taking the moral high ground, *Home Chat* extols the virtues of looking on the bright side of things, even if 'her pastry may not always be just right, and she may occasionally burn her bread, and forget to replace missing buttons …' Continuing in a somewhat sugary vein it presents this verse:

> *Sweet is the smile of Home; the mutual look*
> *Where hearts are of each other sure;*
> *Sweet all the joys that crowd the household nook,*
> *The haunt of all affections sure.*

By the mid-19th century advice for women was becoming ever more profuse. A welcome note of moderation could be found, however, as in this paragraph from *The Practical Housewife* of 1860: 'A woman who worries all within her reach by her ultra-housewifery, who damps one down with soap and water, poisons one with furniture polish, takes away one's appetite by the trouble there is about cooking the simplest thing, and fidgets one by over-done preciseness and cleanliness is almost as much to be avoided as a downright sluggard, or the veriest simpleton.'

According to the critic, author and artist John Ruskin the prime duty of an Englishwoman towards her house was 'its order, comfort and loveliness'.

A PRETTY DECORATION WILL GRACE ANY TABLE

As long as it is not so large that it obscures the view – and the conversation – across the table and does not make it too cluttered.

Extolling the merits of a popular design of vase named 'The Excelsior', *Cassell's Household Guide* says that 'of all crystal table decorations, it is best fitted for a centrepiece. It consists of a very large, flat, circular dish of glass, from the centre of which rises a slender shaft of the same bright material surmounted by cornucopia. Round the side of the dish three other cornucopias are arranged; these as well as the dish, are dressed with flowers, fern fronds, grasses, roses, stephanotis, geranium, passion flowers, fuchsias, &c.' For a less formal effect, small vases of flowers or bowls of fruit are pretty – just a few simple flowers can transform a table.

Good containers are also adaptable. 'A pretty table decoration in the strawberry season,' the *Guide* says, is a glass dish 'piled high with strawberries … For a strawberry party a small stand of such a kind is placed to every guest, a tiny glass bowl of sifted sugar, and a little jug of cream.' Such a presentation would still be delightful.

Candles on the table must be chosen with care. If they alone are lighting the table then there needs to be one per person. Larger candlesticks and candelabra need to be high enough so as to prevent light shining directly into diners' eyes, and chosen in proportion to the size of the table.

Details deserving the 'utmost attention', says Enquire Within *on the matter of laying the table, are: 'The whiteness of the table-cloth, the clearness of the glass, the polish of plate, and the judicious distribution of ornamental groups of fruits and flowers.'*

NAPKINS SHOULD ALWAYS BE NEATLY FOLDED

That is, when set on the table before they are used. Polite guests will fold their napkins at the end of a meal rather than just throwing them loosely on to the table.

Napkin or serviette? According to the dictates of 'U and non-U', begun in 1954 by Professor Alan S.C. Ross and taken up by the author Nancy Mitford in her 1956 book *Noblesse Oblige*, 'napkin' is the proper usage.

Simple is best when it comes to the presentation of the napkin. 'Very fancy foldings,' says Emily Post in her etiquette classic of 1922, 'are not in good taste, but if the napkin is very large, the sides are folded in so as to make a flattened roll ... If they have a corner monogram they may also be folded in half, and the two long ends folded together.'

In formal company the napkin is used on the lap, not tucked into the collar, although in medieval times guests would throw their napkins over their shoulders. However napkins and tablecloths were, at this time, the preserve of the rich and successful. Centuries earlier the Romans had introduced *sudaria* or 'handkerchiefs' – pieces of fabric used primarily for wiping the brow, not the hands, during a meal. Larger cloths known as *mappae* were spread over the couches on which diners reclined, and were used to blot the lips and wipe the hands after eating.

Currently, napkin rings are out of fashion, but they are still worth collecting. They were a Victorian invention, first made in silver in sets engraved with numbers or initials. By the mid-1800s they had become popular as christening gifts, most valuable when individually crafted.

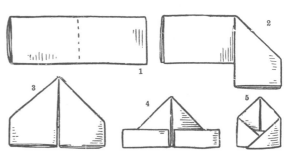

A SIDEBOARD WILL GREATLY RELIEVE A CROWDED TABLE

It will, if you are lucky enough to have the space for one in your dining room. The sideboard can be used for carving, for extra dishes waiting to be served, and for wines.

The sideboard began life as a simple table or 'slab', often with a marble top, on which food was cut and from which it was served. By the late 18th century, however, it had come to combine a counter, cupboards, a cutlery drawer and a cellaret – that is, a drawer for holding bottles and decanters, in effect a miniature cellar. Some of the finest sideboards were designed by Hepplewhite and Sheraton, typically of mahogany inlaid with boxwood (see box). By the mid-1800s the pedestal sideboard, looking more like a desk, had become popular. Although smaller in size it was in fact a heavier piece of furniture, in keeping with the tastes of the age.

The use of the sideboard reflects the changes in the way that food was served. Until the early 19th century, when the Russian Prince Alexander Kurakin introduced his *service à la Russe* to France, many dishes were laid on the table at once, with 'reserves' on the sideboard. But with the new style of service dishes were served one at a time, making the sideboard not just useful but almost essential for large family meals as well as entertaining. As *Cassell's Household Guide* observed in the 1880s: 'Vegetables are never served except from the sideboard, at any except strictly family dinners.'

As well as accommodating food, a sideboard lends itself to decorations, some of them practical. 'Among useful decorations for the sideboard, some of the prettiest I have seen,' says Mrs Caddy in her *Household Organization* of 1877, 'are the Venetian bottles for holding oil and vinegar.' These had curved necks and were fixed in a glass so that they curved across each other. 'This simple yet ingenious contrivance,' she adds, 'is far prettier than our somewhat vulgar cruet-stand.'

The original credenza was a side table on which food was placed for tasting (in case of possible poisoning) before being served to a monarch, pope, lord or some other high-born person. The term comes from the Latin *credentia*, meaning 'trust'; it now describes a flat-topped cupboard of table height, which is used as a display cabinet or bookcase.

CLASSIC SIDEBOARDS COMPARED

Though similar in many ways, the sideboards of Hepplewhite and Sheraton display marked differences:

HEPPLEWHITE – Front with serpentine or reverse curves, six square legs often ending in a spade foot and ornamented with vertical patterns of husks. Made of mahogany, either solid or veneered. Inlay either a narrow line border or a fan or wreath design. Sometimes decorated with carvings of ribbons, flowers, husks, urns and wheat ears.

SHERATON – Front with bold swelling curves and often a straight centre. Six rounded slender, tapering legs, usually reeded. Made of mahogany with ornamental inlay in patterns including medallions, fans, vases and shells. Carving, if present, is in simple Greek patterns. More elaborate interiors with cupboards for wine bottles and racks for glasses and plates. Usually with a brass railing, often elaborately ornamented and incorporating candlesticks at the back of the top.

NO COURSE DURING A DINNER REQUIRES GREATER ATTENTION THAN THE SOUP

The reason being, says *Home Chat* of 16 May 1896, that 'it foreshadows, as it were, what the guests may expect from the liberality of their host and the competence of his cook.' Equally, polite guests will know how to eat their soup with decorum.

Every cook – and every cookbook – has plenty to say about soup. Mrs Beeton, for instance, concludes: 'The principle art in composing good rich soup, is to proportion the several ingredients that the flavour of one shall not predominate over another, and that all the articles of which it is composed shall form an agreeable whole.' She also advises: 'Stock being the basis of all meat soups, and, also, of all the principal sauces, it is essential to the success of these culinary operations.' Certainly a good stock remains the basis of excellence in soup making.

When it forms part of a meal, soup is invariably served as the first course and may be ladled into bowls from a tureen. Old-fashioned etiquette on the subject is stern: you should sup your soup from the side, not the end, of the spoon and tip the soup plate away from, not towards you. Slurping is never acceptable, but if you are presented with a soup cup with two handles then it is polite to lift and drink direct from the cup.

There are few better homages to soup than the Mock Turtle's song from Lewis Carroll's *Alice Through the Looking Glass*:
'Beautiful soup, so rich and green,
Waiting in a hot tureen!
Who for such dainties would not
    stoop?
Soup of the evening, beautiful Soup!
Soup of the evening, beautiful Soup!
Beau--ootiful Soo--oop! Beau--
    ootiful Soo--oop!
Soo--oop of the e--e--evening,
Beautiful, beautiful Soup.'

At an Evening Party, Chess and All Unsociable Games Should Be Avoided

For evening entertainment, dancing and card games were once both expected and acceptable. Aside from the rules of games, there were many other rules that any host and hostess – and their guests – would be expected to observe.

The modern standard pack of cards arrived in Britain from France in 1470 and was the first to contain a queen. Before this, court cards included a king and a deputy king.

For the host and hostess, *Enquire Within* includes the following pointers: 'Be cordial when serving refreshments but not importunate ... there should be a table for cards and two packs of cards placed upon each table ... The host and hostess should look after their guests and not confine their attentions. They should in fact attend chiefly to those who are least known in the room ... Pay respectful attention to elderly persons.'

For the guest, typical rules of party etiquette to be followed were: 'Upon entering, first address the lady of the house; and after her, the nearest acquaintances you may recognise in the room ... If you introduce a friend, make him acquainted with the names of the chief persons present. But first present him to the lady of the house, and to the host ... If there are more dancers than the room will accommodate, do not join every dance ... Avoid political and religious discussions. If you have a hobby, keep it to yourself.'

Card games popular at evening entertainments would have included whist, cribbage, and vingt-et-un. Bezique, Napoleon and poker were extremely popular and betting was acceptable, though for politely low stakes. Spectators were also required to behave properly. 'Lookers on at the game [of whist],' says *Cassell's Household Guide*, 'are not allowed to make any remarks; but they may be appealed to as referees to decide a doubtful question, as to who played a particular card, what is the law of the game upon a certain point and similar matters.'

THE HOSTESS SHOULD NEVER APOLOGISE FOR THE SIMPLICITY OF THE MEAL

It is more important to serve your guests with excellent ingredients, well cooked, and to be attentive to everyone's needs. If food, however simple, is home grown or locally sourced, then so much the better.

The table, it is said, is the first principle of hospitality, and this remains true whether you are serving a simple supper for a few friends or holding an elaborate dinner party.

The wonderful selection of ingredients now available from all over the world, plus the plethora of recipes and cookery programmes, makes it easy for even the most inexperienced cook to make a tasty, easy meal for friends. A simple dinner from before the 1960s, when 'exotic' vegetables such as aubergines (eggplant), avocados and peppers became readily available, and people's tastes became more adventurous, would have seemed dull and bland indeed to the modern palate.

Instructing hostesses on what to serve, the 1958 guide *Teach Yourself Etiquette and Good Manners* suggests the following somewhat mundane menus for dinners in different seasons. As was the vogue at the time, a savoury, rather than a cheeseboard, was served after the pudding. The 'Dessert' that followed it could in fact have been cheese, or fresh fruit and nuts. Petits fours would be offered to accompany the coffee.

SUMMER	WINTER
Grape Fruit.	Clear Soup (Julienne).
Boiled Salmon (Hollandaise Sauce).	Boiled Turbot (Shrimp Sauce).
Roast Chicken (Bread Sauce) with Peas and Potatoes.	Lamb Cutlets (Mint Sauce). Brussels Sprouts, Potatoes.
Fruit Salad and/or Meringues.	Trifle and/or Wine Jelly.
Mushrooms on Toast.	Cheese Soufflé.
Dessert.	Dessert.
Coffee.	Coffee.

Whatever they are planning to cook, and however simple, thoughtful hosts will ask their guests in advance whether there are any foods they are unable to eat and plan the menu accordingly. It may, as a result, be necessary to provide alternatives for vegetarians, but polite guests will be graceful in refusing any food they are unable to consume.

At table, every guest needs elbow room

Ideally yes, but in convivial company and with family and good friends, it is unlikely that guests will mind sitting close together.

> It is said that the chief maxim for dining in comfort is to have what you want when you want it.

'Whatever the desired number of guests may be,' says *Cassell's Household Guide*, 'the invitations should be limited by the size of the dining room. At least sixteen inches of room should be allotted to each individual … For all purposes of agreeable conversation,' it continues, 'eight persons are as many as should be invited at a time. It is no compliment to crowd guests at table … Less on the table and fewer to consume is better taste.'

For an elaborate dinner with many courses, space at each setting is vital for items of cutlery and different glasses for each wine for, as the *Guide* rightly says, 'No one would think of pouring sherry into a champagne glass, or *vice versa*.'

At a typical late Victorian dinner, however, vegetable dishes would not be put on the table but served from the sideboard. In traditional formal dining, it was impractical to load up the table with all the many utensils, plates, bowls and glasses the guests would need to navigate a large meal of six or more courses. Thus the table was set with only enough silverware for the first few courses (up to six pieces, with no more than three on each side), and up to three appropriate glasses arranged in a triangular shape above and to the right of the dinner plate. Following similar lines today, even on a smaller scale, still makes perfect sense.

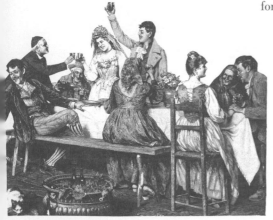

It Greatly Facilitates the Ease of Seating Company When the Names of Guests Are Placed by Their Plates

Not only does it greatly assist the business of seating guests,
it allows hosts to work out their seating plan well in advance.
Pretty hand-crafted place cards make nice keepsakes for guests.

In more formal times, correct seating, according to the strict rules of etiquette,
was vital in polite society. The master of the house would be seated at the head
of the table, generally in the position farthest from the door, with his back to the
sideboard. To his right would be seated the most distinguished lady, whom he would
accompany into dinner. The lady of the house would sit at the opposite end, with
the most distinguished gentleman to her right, the second most important to her left.
Other guests would be placed, men and women alternately, in between.

Etiquette guru Emily Post was most particular about place cards, specifying
that they should be 'plain, about an inch and a half high by two inches long'. She
was vehemently against 'fancy cards', considering them suitable only for Christmas
and birthdays and unsuitable for a formal table. The accomplished hostess would
write her own cards in an impeccable hand. At a formal dinner, even in the home,
it would also have been usual for menu cards to be placed on the table. *Enquire
Within* says that the menu should be 'neatly inscribed upon small tablets, and

distributed about the table, that the diners know what there is to come'.

When places are found, true gentlemen will 'seat' their female neighbours at table by pulling out their chairs and helping them to get close to the table before the hostess sits down. Polite guests will converse equally with those seated to either side of them, and to those seated opposite, although it is no longer the case that the men should feel obliged to initiate topics of conversation

A FINGER BOWL SHOULD ACCOMPANY ANY FOOD EATEN WITH THE HANDS

Known in Victorian times as the finger glass, the finger bowl remains essential for cleaning up after eating a lobster or mussels, or vegetables such as asparagus and globe artichokes.

Victorian finger glasses were usually brightly coloured in purple, green, pink and blue and, unless needed earlier, were brought to the table with dessert. 'However neatly a person may eat,' says *Cassell's Household Guide*, 'sugary sweets and juicy fruits will leave a trace on the fingertips: not to mention asparagus, smelts, peach or apple fritters, or gingerbread cakes if handled when eaten, which is perfectly orthodox. Shrimps and other crustaceans, which also are allowed to come into contact with the finger and thumb, betray the presence of saline elements.'

When presented, the water in a finger bowl is usually warm. Today, slices of lemon are the usual addition for use with savoury items, although these were derided by Emily Post as 'never seen in a well-appointed home in an after-dessert finger bowl'. For the latter, rose petals, violets or eau de cologne were *de rigueur*.

'After broiled lobster,' she concedes, 'at an informal or family dinner, lemon in *hot* water (or soapy water) is excellent.'

It is permissible to eat with your hands because, in the words of the proverb, 'Fingers were made before forks and hands before knives.'

FINGER FOOD

The rules of etiquette are most particular when it comes to foods eaten in the hands. After using a finger bowl, wipe your hands on your napkin.

GLOBE ARTICHOKES – leaves are pulled off one at a time, dipped in a sauce and bitten off. The centre, after the bristly choke has been removed, is eaten with a knife and fork.

ASPARAGUS – can be picked up in the fingers and eaten tip first. Any hard end pieces should be returned to the plate.

CORN ON THE COB – messy even if holders are supplied for the ends. Pick up the cob and bite off the kernels. The good host will provide toothpicks as well as finger bowls.

PRAWNS AND SHRIMPS – pull off the head then loosen the shell on the underside, between the legs, and pull sharply on the tail.

MUSSELS – use an empty shell or a fork to remove the meat.

LOBSTER – use a lobster fork to prize out the meat. The claws will need to be broken with crackers.

CHEESE IS THE PEARL OF DESSERTS

And as such is a perfect course to serve at a luncheon or dinner. A respectful host will offer cheese before the sweet dessert – in the French style – as well as after it.

The well-prepared cheeseboard will present a good variety of cheeses, both hard and soft and of varied strength, including some blue cheese. The cheese should be served at room temperature, not direct from the refrigerator, so as to allow its flavour to be appreciated to the full. When selecting cheese for guests, look for local specialities and make sure it smells wholesome and shows no sign of sweating. Any good delicatessen or market stallholder will allow you to taste before you buy.

Knowing how to cut a cheese is essential etiquette. A small round cheese such as a Camembert or a *chèvre* should be cut from the centre in a wedge. A piece from a larger cheese such as Brie or Stilton, already in a wedge shape, should be cut along the side, preserving its sharp end or 'nose'. The old-fashioned cheese knife has a forked tip for removing the cut cheese to your plate. Modern versions are plainer and often sold in sets of three so that the cutting of one cheese does not 'taint' another. Biscuits are served, but purists will never serve butter with cheese, so as not to mask the purity of its flavour.

In Britain, cheese was considered 'peasant fare' until the 16th century, when varieties such as Stilton, Cheshire and Banbury were beginning to be appreciated. By this time cheese from Holland, Italy and France was already being imported – and enjoyed. In Scotland cheese was so valued that in 1573 a law was passed banning its export.

An old way of keeping cheese fresh, moist and free of mildew was to place two sugar lumps on the cheese dish before covering it.

A dessert without cheese is, it is said, like a beautiful woman who has lost an eye.

Pure water is better than foul wine

Although good hosts will serve both water and excellent wines to their guests. Glasses for water should be placed on the table alongside wine glasses.

The most highly regarded of all drinking water has long been naturally sparkling mineral water containing bubbles of carbon dioxide. While the Romans appreciated its qualities, they would bathe in it rather than drink it, and only in the Middle Ages did its excellence as a drink come to be appreciated. Today, naturally effervescent waters such as Apollinaris, Badoit and Vichy are valuable commercial brands. The first artificially carbonated water was produced in 1767 by the English chemist Joseph Priestley and was on sale from 1781. In the USA, plain and flavoured carbonated water was sold by the millions of gallons once soda fountains became universally popular in the 1820s.

Today in the developed world we can rely on the purity of our tap water and drink it safely, even if we are not partial to the taste of the chlorine and other chemicals used to free it from infective micro-organisms. In 1883 the guide *Our Homes* provided readers with a set of pointers for judging the purity of water, stating that it should be colourless or bluish; bright and sparkling; have a pleasant, sparkling taste; be devoid of any smell; and soft to the touch. 'Although it is impossible,' adds a caveat, 'to guarantee that such water shall be absolutely free from danger, yet we may go so far as to say that no water can be a first-class water where those qualities are absent.'

The ideal glasses for serving water are tumblers, named from the fact that they originally had rounded bases and therefore could not be set down without tumbling over. They may be plain or of cut glass. Good antique tumblers, such as those sold in the 17th century by the London glass merchant John Greene, are highly prized collectors' items.

FLAVOURED WATERS

Slices of lemon, orange or lime in a water jug will impart a tangy, refreshing flavour. Fruit waters were popular beverages in the 19th century. Examples include:

RASPBERRY WATER – made by adding a few drops of raspberry vinegar to water.

APPLE WATER – made by slicing and steeping two large apples in boiling water, which is then strained and cooled.

CURRANT WATER – two teaspoons of redcurrant jelly put into a tumbler of water with a pinch of tartaric acid.

PRUNE WATER – a handful of prunes and half a lemon simmered in water, strained and cooled.

SERVE WHITE WINE WITH FISH, RED WITH MEAT

A helpful guide, not an unbreakable golden rule. The secret of success in matching wine with food is to think of the wine as another ingredient.

Following this logic, it makes perfect sense to enjoy an acidic wine, such as a Sauvignon, Riesling or Muscadet, with fish in the same way that you would squeeze lemon juice over it. Similarly, a rich, full-bodied wine such as a Shiraz or Merlot is an excellent accompaniment to beef or game, while the acidity of a Cabernet Sauvignon makes it a perfect foil for the richness of roast pork.

If red is your preferred colour but fish your food, it is best to avoid wines high in tannins, such as vintage Burgundies. But there are many good, light, acidic reds to choose from, including young Riojas, Chiantis and Valpolicellas. For a white that will complement a meat dish, body is what is required. You will get it from wines such as Chardonnay, Pinot Gris and Semillon.

In former times, wine was routinely bought by the cask and bottled at home. The butler, who was in charge of the cellar, was responsible for this task, and for laying down the wine.

Some 800 years ago the School of Salerno in Italy, a prestigious academic centre in the Middle Ages, defined the qualities of the perfect wine in terms of five Fs: *fortia, formosa, fragrantia, frigida, frisca,* which translates as 'strength, beauty, fragrance, coolness, freshness'.

In the grand houses and restaurants of 1890s America, champagne was often served throughout a meal 'with no other adjunct but bottled waters'. But, as the social adviser Constance Cary Harrison remarked, 'How infinitely more welcome to the habitual diner-out is a glass of good claret than indifferent champagne!'

After dinner, the port should be passed

And it must be passed correctly. The tradition of drinking port in Britain goes back to the 17th and 18th centuries, when wars against France, beside stifling the Anglo-French wine trade, made it unpatriotic to drink French wines at dinner.

The traditions of port drinking after dinner, still firmly upheld in officers' messes, Oxbridge colleges and London clubs, go back to old naval practices. First to take the port decanter is the host, who pours a glass for the guest on his right. He then passes the decanter to the left. Each person fills their own glass, then hands the decanter to the next person to the left. When the decanter reaches the host he pours himself a glass. As the decanter is passed it should never touch the table or be lifted over a glass.

A port wine mark is the name for a type of birthmark that is the colour of the drink.

Port is wine from the Douro Valley in northern Portugal fortified with brandy. The spirit acts to stop the fermentation, leaving a residual sugar in the wine and boosting its alcohol content. Fortification was introduced by English importers to enable the wine to withstand the long sea journey from Portugal, but it established a British taste for fortified wine. Wines in the same style are made in other countries but may not bear the *appellation* of port, first granted in 1756.

Port may be matured in wooden casks or barrels or in sealed glass bottles. Barrel ports include tawny, colheita and garrafiera, while ruby, pink and white ports are bottle aged. A vintage port is one made entirely from grapes harvested in an officially declared vintage year. Vintage ports are aged in barrels for up to two and a half years before being bottled, and acquire their complex flavour from the slow decomposition of grape solids in the bottle.

THE QUESTION OF LIGHTING THE TABLE IS VERY IMPORTANT

Because it needs to be subtle and flattering to both food and guests but not so dim as to obscure what is on people's plates. Candles are always a good choice.

If a candle will not fit into a candlestick, dip the end into hot water to soften the wax and set it in place while the wax is still warm.

'A dining-room chandelier seldom suffices to give sufficient light for a festive occasion.' So says *Cassell's Household Guide*, pronouncing on the weighty matter of dinner parties. 'Branched candelabra, containing wax candles, are the most suitable,' it advises, 'and lamps are the least convenient …

Lamps are very useful for lighting the mantelpieces and sideboards of a dining-room, where they aid in producing a generally diffused light, so desirable at a dinner-party.' An alternative, still most acceptable now, is 'several candlesticks judiciously placed among the ornaments and at the corners of the table'.

As to choice of candlestick, the guide recommends that small plain glass candlesticks, 'such as are sometimes used for lighting a pianoforte, are in good taste at an unpretending dinner, where glass, and not silver is the principal feature in the service.'

Silver candlesticks are, of course, most desirable. The earliest ones, dating from the 1660s, were made of sheet silver, usually with an octagonal base and fluted stem. By the late 18th century silver candlesticks shaped like Doric or Corinthian columns were popular, and designs of this kind are reproduced to this day.

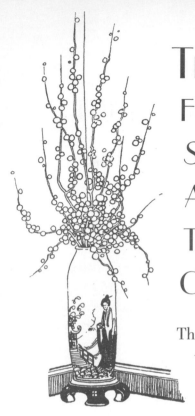

THE VASES IN WHICH FLOWERS ARE PLACED SHOULD BE IN ACCORDANCE WITH THE SURROUNDINGS OF THE SITUATION

There is no better way to make a home welcoming to family and guests than with flowers judiciously placed and, of course, arranged in an appropriate vase.

Cassell's Household Guide is specific in its recommendations: '… terra-cotta, Wedgewood [*sic*], and majolica ware,' it says, 'are most suitable for halls and staircases; whilst delicate biscuit ware, fine porcelain and Bohemian glass, are better adapted to drawing-rooms, and clear crystal to the dinner-table.'

Flowers arranged in the home are always most effective when appropriate to the season, even if they have not been cut from the garden. The chrysanthemum may be, in the words of Muriel Spark's Miss Jean Brodie, 'a serviceable flower', but it naturally blooms in autumn and is not fitting for a spring or summer arrangement, even if it is available then. Scented flowers of any kind are excellent for a hallway, where their perfume will waft through the house, but the *Household Guide* warns that some floral scents can easily become oppressive: 'Such shrubs as lilac and syringa, when in bloom, are inadmissible in sitting-rooms. The same may be said of wall-flowers. Few persons can bear the odour of such plants without inconvenience in a close apartment …'

The vase is one of the most ancient forms of artefact, but apart from antiquities and pieces imported from China and other places in the Orient, vases were virtually unknown in Europe until the 17th century, when there was a veritable explosion of decorative forms. By the middle of the 18th century, when flower arrangements became accepted as a form of interior design, superb vases were being made by factories such as Wedgwood, Meissen and Sèvres, although most would be considered far too precious for such a mundane purpose as floral display.

FLOWERS FOR THE HOME

Additional tips for the suitability of floral arrangements from *Cassell's Household Guide*:

Whatever their size, flower vases should be shaped so as to contain plenty of water.

Flower arrangements are not suitable for the bedroom.

Sideboard decorations are most appropriate when composed of flowers growing in their own pots.

Highly scented flowers do not mix well with the aromas of food and therefore should be kept out of the dining room.

Roses are best when seen together in large quantities, but require foliage to show them to best advantage.

A BED FOR A GUEST SHOULD BE WELL AIRED

It should also be warm and comfortable. In cold weather guests may appreciate a hot water bottle or – if you have one – an electric blanket.

To test whether a bed is dry and well aired you should, says *Enquire Within*, 'introduce a drinking glass between the sheets for a minute or two, just when the warming-pan is taken out; if the bed be dry there will only be a slight cloudy appearance on the glass, but if not, the damp of the bed will collect in and on the glass and assume the form of drops – a warning of danger.'

The warming pan, rather like a large enclosed frying pan on a long handle, came into use in the 16th century. Usually made of copper (a good conductor of heat), it was filled with hot coals before being inserted into the bed. To get the coals hot quickly, maids knew the trick of throwing a handful of salt on them. The warming pan became obsolete with the advent of the hot water bottle, originally made of glass or earthenware and subsequently of rubber.

The first electric blankets appeared in the early 1900s but they were big, bulky and dangerous, causing numerous bed fires. However, by the 1920s they were favoured for tuberculosis patients who, because of the supposed advantages of fresh air, routinely slept outdoors. Safety was greatly improved with the introduction of thermostatic controls from 1936.

> After rising in the morning the bedding should be turned back for a while to air. Jane Panton, author of *From Kitchen to Garret*, was convinced that she had 'never yet found in all my experience a servant who can really and truly be trusted to properly air the bed', and that this should be a job for the lady of the house.

PLACE BOOKS IN READINESS FOR THE SLEEPLESS GUEST

Because the good hostess should be sure to adapt a spare bedroom to the comfort and convenience of her visitors. Guests will need proper lighting to read by and space in drawers and cupboards in which to place their clothes.

As to the choice of reading matter, a good selection of paperbacks, including some classics, will suffice, but it is much more engaging to include books relating to local and family interests and, if possible, to cater for the preferences of individual guests. A selection of magazines will also prove diverting.

'Every hostess,' says Emily Post, 'should be obliged to try her guest room by spending at least one night in it herself. If she does not do this she should at least check the facilities thoroughly.' She goes on to list all the essentials needed, including plenty of bath towels, face towels, bathmat, and new cakes of soap for the bath and wash basin. Two pillows per guest and extra blankets in case of cold are also vital.

In the country the perfect hostess will keep a store of umbrellas and Wellington boots in different sizes for use by her guests.

Regarding lighting, bedside lights are a must: one on each side of a double bed or one beside each twin bed. In the cupboards, wooden, rather than plastic, coat hangers are always preferable and the selection should include some designed to accommodate skirts and trousers as well as coats, dresses and shirts. It is a sign of her times that Emily Post also says that there should be 'cigarettes, matches, and ash trays on the tables'.

TAKING TEA IS A MOST GENTEEL AMUSEMENT

Except for children's birthdays, the tea party is no longer much in fashion. In times past, however, entertaining friends for tea was a social highlight, and often an occasion of great formality.

From the time tea was first drunk in England, sharing it with a convivial group became an excuse for good food and conversation. It was one of the commodities imported by the British East India Company, under the terms of the royal charter granted to it by Elizabeth I in 1600, but the fashion for tea drinking dates from the 1660s and is attributed to Catherine of Braganza, the Portuguese wife of Charles II, who introduced the habit to the English court. She instigated formal teas – as main meals – for which guests had to dress as formally as they would for a state dinner.

Afternoon tea as a social event was introduced in England by Anna, the seventh Duchess of Bedford, in 1840. As the fashionable hour for dinner grew ever later – from mid-afternoon in the 18th century to after eight in the evening in the early 19th century – it left a long period without refreshment. The Duchess filled the gap by requesting a tray of tea, with bread and butter and cake, to be brought to her room during the late afternoon. As this became a habit, she began inviting friends to join her. By the 1880s, the wardrobes of well-to-do society women included long gowns, gloves and hats especially for afternoon tea.

For a formal tea party a hostess would send out invitations at least ten days before the event. Before the day she would make sure that all her silver was polished. The ritual of serving tea, usually taken in the drawing room

According to legend, tea was discovered by the Chinese Emperor Shen Nung in 2737 BC, when some leaves from a tea plant accidentally fell into some water that he was heating. The resulting drink refreshed and revitalized him, and tea drinking was 'invented'.

between four and five o'clock, was highly stylized, and provided the hostess with a chance to show off her best china and linens and to serve delicate sandwiches and cakes.

While some people insist that tea's flavour is best (and good manners most satisfied) when milk is poured into the teacup first, others maintain that it is only possible to get the strength exactly right if the milk is added afterwards. The British were the first to add milk to tea, probably to soften its bitter tannins. Putting the milk in first could also have helped to prevent fine Chinese porcelain tea bowls from cracking with the heat of the tea.

The taste for sweetening tea with sugar began in Europe in the 17th century and was enthusiastically adopted by the British, who not only added sugar tongs, spoons and bowls to their tea sets but exported their enthusiasm to North America.

CHAPTER 6

THE WELL-MANAGED HOME

'Frugality and economy are home virtues without which no household can prosper.' This mantra from Samuel Johnson, quoted by Mrs Beeton in her advice on achieving a well-managed home, is as true today as it was more than two centuries ago. When budgets are balanced and nothing wasted a home is likely to run smoothly and economically. The good housewife of the past would keep weekly accounts, recording expenditure to the last farthing on everything from servants' wages to bills from the butcher, fishmonger and grocer.

Home makers of previous generations were adept at making the best use of every possible item – a virtue now becoming ever more appreciated. They would remove the buttons from old shirts before discarding them, cut up old clothes and sheets for dusters, and save and dry orange peel and potato peelings for making firelighters. Garden flowers were cut and dried for winter arrangements, and both children and adults would entertain themselves by making papier mâché objects from waste paper.

Needlework of all kinds was once central to home making. As well as dressmaking, which was once, for many women, the only affordable way of keeping up with fashion, women and girls would embroider and knit, and make rugs and tapestries. Torn clothes would be patched and darned. The art of the needle, dubbed 'a woman's most useful weapon', was especially relevant in wartime, when men, women and children were all encouraged to knit items of clothing for the nation's soldiers, sailors and airmen.

PATCH BY PATCH IS GOOD HOUSEWIFERY

Meaning that it is better to patch a worn garment than to throw it away. When used for making quilts, patchwork is not only skilful but also highly decorative.

Oddments of lace can easily be joined to make table mats or individual pieces of work that can be mounted and framed.

For a neat patch on plain fabric, the generous patch of new material is placed on the wrong side and held in place with a simple fell stitch or a neat herringbone. On the right side, the torn fabric is neatened, and the edges fell stitched into place, with the raw edges turned under if necessary to prevent them from fraying. 'For print and other patterned materials,' says Thérèse de Dillmont in her *Encyclopedia of Needlework*, 'the patch must be arranged on the right side of the garment, the edges being turned in accordingly. Tack it on so that the stripes or pattern exactly match in all directions, then seam it on …'

101 Things to Do in Wartime, published in 1940 with frugality at a premium, advises that 'Patchwork provides an excellent and economical way of putting to practical use all kinds of odd and ends of plain and coloured fabrics. Counterpanes, cushion covers, table cloths and innumerable small articles, by the exercise of patience and a little ingenuity, can be made highly decorative.'

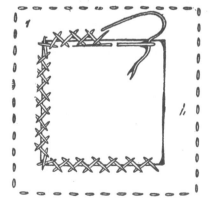

The first patchwork quilts were created from scraps, but fabrics are now bought in 'fat quarters', so called because each linear metre or yard is cut into four squares rather than narrow strips. Quilting was an essential art for early American settlers, and quilts were common wedding gifts, in designs with such evocative names as 'Hearts and Gizzards' and 'True Lover's Knot'.

A woman's most useful weapon is her needle

Certainly as far as home economy is
concerned, whether it is used for
sewing, embroidery or mending.
Steel needles have been in use
since the Middle Ages. As well as
the correct needle, a thimble is essential to hand work.

A needle needs to be chosen to fit the thread you are sewing with and the fabric
on which you are working. The more delicate these are, the smaller the eye should
be and the slimmer the needle. An old test of the quality of a batch of needles is
to try breaking one between your fingers. If the steel is well tempered it will be
hard to break and, if it does, will snap cleanly. To save needles from rusting it was
customary to keep them in a needle case made by sewing several small squares of
flannel together to make a kind of book.

A good sewing needle should never be bent – it will make ugly, irregular
stitches. It should also be a little thicker than the thread you are using, to allow
the latter to pass easily through the fabric. For plain
sewing short needles are perfectly acceptable, but
for any other kind of sewing longer needles are
best. If threading a needle is a problem, an old trick
is to hold it in front of a piece of white paper, so
sharpening the outline of the eye.

The original thimbles, which were worn on
both fingers and thumbs, were made of leather, and
the word is derived from the Old English *thymel*,
meaning 'thumb stall'. The earliest known metal
thimble, made of bronze and found at Pompeii, dates
to the first century ad.

> To break a needle while sewing is
> said to bring good luck to the person
> who will wear the garment being
> sewn. To drop a needle and pick
> it up yourself is to encourage ill
> fortune, but a dropped needle that
> lands on its point is a harbinger of
> good times.

Always darn on the wrong side of your stuff

Definitely the place to start. The regular darning of woollen socks undertaken by our grannies might no longer be necessary, but darning is a skill worth acquiring to prolong the looks and lives of precious garments, whether they have been damaged by moths or simply by wear and tear.

The aim of good darning is to imitate, across a hole, the weave or structure of the original fabric. This means that your choice of darning thread is critical: it should match the thread of the fabric to be repaired in both weight and colour. Using a darning needle – a long, heavy needle with a large eye and a blunt tip – this, step by step, is the way to proceed:

1. Begin on a sound area well away from the edges of the hole, running the needle in and out of the fabric in a straight line.
2. Return as close as possible to the first line, going in and out of the fabric as if weaving.
3. At the end of each line, do not pull the thread tight but leave a small loop to allow for any shrinkage of the darning thread.
4. Continue darning up and down, leaving the parallel threads lying across the hole, until well outside the hole on the far side and on sound stuff once more.
5. Then repeat the stitching at right angles, weaving alternately under and over the original lines of stitches until the hole is completely filled.

For sock darning, a wooden 'mushroom', which held the heel or toe taut, was a useful accessory. *The Household Encyclopedia* of the 1930s makes this suggestion:

Pattern darning is a type of embroidery in which parallel rows of straight stitches of different lengths are used to create a geometric design.

'To darn neatly stretch the material over the glass bulb of an electric torch and then hold the sides with an elastic band. If necessary switch on the light for a few moments and it will be quite clear which way the strands of material go. Delicate materials can be repaired easily in this way.'

HOME DRESSMAKING IS A HELP FOR THOSE WITH SMALL PURSES

A saying that is not as true today as it was in the past, when ready-made clothes were expensive and fabric relatively cheap. But for those skilled with pattern and needle, dressmaking can still be worthwhile, particularly for wedding, formal and evening wear.

Two 19th-century advances combined to ensure the popularity of home dressmaking, making fashion available beyond the province of the wealthy. The first was the sewing machine, made possible by the invention of the lockstitch by the American Walter Hunt in 1834, and the patenting of the up-and-down needle movement by Boston mechanic Elias Howe in 1846; the home sewing machine was successfully produced by Isaac Singer in England in 1856. The second was the paper pattern. 'Dressmaking without

a paper pattern for a guide,' said the women's magazine *Home Chat* in 1896, 'too often proves a waste of money ...' By this time, over 30 years after the American tailor Ebenezer Butterick produced the first such pattern – for a man's shirt – in 1863, it was the accepted way of working. Without such a pattern, said the article, 'The various parts show a provoking tendency not to fit in with each other, and when, after sundry unpickings, the garment is finished, it is found to be almost unwearable because it looks so very unlike the work of a professional.' *Home Chat* offered patterns to its readers, priding itself on quality and reliability. A standard flat pattern cost sixpence halfpenny, one based on a reader's measurements two shillings and a penny. 'All patterns,' said the small print, ' are cut on the best possible French method to all the measures and requirements of the wearer and each is, therefore, an individual work.'

TIPS FOR HOME DRESSMAKERS

Under the heading 'every girl her own dressmaker', *Household Science* of 1889 included dozens of tips and hints, such as:

To begin with, attempt the patterns of only simple garments.

It is easiest for amateurs to make up plain materials for stripes and patterns require to be matched ... All sprigs and flowers must be made to run upwards.

Learn to cut out exactly.

Know the right and wrong sides of all fabrics.

Be careful not to pucker the seams in putting the material together.

Mark where the seams are to be, then tack [the garment] together and try it on.

The art of knitting requires only that the fingers should be properly used

Maybe easier said than done. The ancient craft of knitting, which is enjoying a 21st-century revival, was greatly encouraged during World War II as a means of helping to clothe the troops.

The oldest real knitted garments – created using two sticks and pulling loops through loops – are socks from Egypt dating to about 1000 AD. Before this, garments were made using a series of multiple knots and loops, in a technique more like crochet. Patterns were incorporated from early on. Often these were symbols to bring blessings or to ward off evil. Cotton, not wool, was the earliest yarn used.

Hand knitting really took off in the Elizabethan period, when everyone wore knitted stockings. Knitting schools were established to provide an income for the poor, with men, not women, as the wage earners. In Scotland, Fair Isle techniques, using wools of many colours, were developed. Such sweaters, made from wool coated with natural lanolin, were resistant to all weathers. The word 'jersey' comes from the Channel Island of that name, which was renowned for its knitwear from the late 16th century.

The technique of knitting has remained basically unchanged since the mid-16th century when the purl stitch was introduced, making ribbing possible. The oldest existing example is in a pair of stockings found in a tomb in Toledo in Spain, from 1562.

By the Victorian era knitting had become a 'parlour art', and women with leisure time to fill would make elaborate lacy items such as bags and baby clothes, stringing beads on to their yarn and knitting them into the fabric.

FOR OUR BOYS

Knitters in wartime were encouraged to make garments for the needy, including evacuees and members of the armed forces. *101 Things to do in Wartime* (1940) declared knitting 'a work of national importance' and included these instructions:

Needed most of all for children are jerseys with collars, socks and stockings, gaiters, suits, skirts and vests ... for the hospitals: bed-jackets, bed socks, etc.

Soldiers and airmen will need pullovers, helmets, gloves, socks and mittens, mufflers.

Sailors will need gloves from 11 to 11½in long, width 4½in, finger length about 3½in, and length of ribbing at wrist 3½in ... Seaboot stockings, length 26in to 28in made from a coarse, hard, natural wool.

Special care should be taken with all knitting intended for use by members of the forces. It is essential that the wool should be of the finest quality.

Kept candles burn the longest

Because keeping a candle hardens the wax, making it melt more slowly and burn more brightly. The simple candlestick is a perfect holder for a candle. Most showy of all are branched candelabra.

The origin of candles is not documented, but cone-shaped candles are depicted on Egyptian tombs dating to about 3000 BC. Since beeswax candles were not made until medieval times, the earliest types would almost certainly have been produced by dipping rushes or flax fibres into animal fat, known as tallow. Often this was left over from cooking and would have had a noxious smell. Ordinary folk continued making candles in this way for centuries, since beeswax, with its clean flame and pure, sweet smell, was beyond their means.

The mass-production of candles began in the 19th century, with the new availability of spermaceti, a wax made by crystallizing oil from sperm whales, which also made them burn longer, and the invention of a machine for the continuous production of moulded candles. By the middle of the century candles were being produced by machine from paraffin wax, which, mixed with stearic acid, created hard, good quality candles affordable by all but the very poorest. Despite this, the frugal Victorian housewife would keep all her candle ends, melt them and use the wax to fill pill boxes. Before the wax set she would add wicks made from waxed cotton thread. The resulting candles were used as nightlights for children.

The earliest candlesticks were carved from pieces of wood. A metal spike on which to impale and fix

> The Romans would burn candles to scare away evil spirits. In Christianity the candle symbolizes Jesus, who was 'the light to lighten the Gentiles'. On Candlemas day, 2 February, it is still the Roman Catholic tradition to bless all the candles that will be used in the church in the year to come.

the candle was a 14th-century invention. The Old Testament describes an early candelabrum when it relates how God commanded Moses to make a candlestick for the tabernacle from hammered gold consisting of a base with a shaft rising out of it and six arms, and with lamps mounted on the six arms and on the central shaft. This persists today in Jewish tradition as the menora.

CANDLE LAW AND LANGUAGE

That candles were precious possessions is reflected in many sayings and superstitions:

CANDLE WORDS
The Devil's candle is a name for the mandrake, said to be given because the plant shines at night.

A Roman candle is a firework that emits coloured balls of fire, interspersed with showers of sparks.

To sell by the candle is a type of auction. A pin is put into a candle and the wick lit. Bidding continues until the pin drops, at which point the last bidder gains the lot.

Something 'not worth the candle' is not worth the trouble of getting involved. It is thought to originate from the theatre, a poor play being unworthy of the money spent on lighting it.

CANDLE SUPERSTITIONS
A candle flame that comes back to life after it has been blown out is an omen of death.

It is lucky to blow all your birthday candles out with one breath – and to make a secret wish as you do so.

If a spark flies off a candle then a letter is in the post to you.

If you light a candle direct from the fire you will end your days in the workhouse.

SAVE ORANGE PEEL TO LIGHT YOUR FIRE

Dried slowly, strips of any citrus peel make excellent fire lighters and smell wonderful – much better than dried potato peelings, which can be used in a similar way. To get a fire going kindling of some kind is essential as well as paper or some other easily combustible substance.

Citrus peel lights easily because it is full of oil. It can be desiccated in a low oven or in an airing cupboard then stored in a dry place until needed. Even with these aids, getting a fire going can be tricky. Mrs Beeton, in her instructions for housemaids, says that a fire is readily made by 'laying a few cinders at the bottom in open order; over this a few pieces of paper, and over that again eight or ten pieces of dry wood; over the wood a course of moderate-sized pieces of coal, taking care to leave hollow spaces between for air at the centre; and taking care to lay the whole well back in the grate, so that the smoke may go up the chimney, and not into the room. This done, fire the paper with a match from below …'

It is said that he who can make a fire well can also end a quarrel.

A fire that is reluctant to burn will be encouraged by air blown from bellows. The best way to use them on a fire that is only partially ignited is to blow on to the coals alongside those that are still alight. After a few blasts you can then blow into the burning fuel, which should be effective in making the flames spread rapidly.

Enquire Within provides a recipe for firelighters consisting of one bushel of sawdust or small coal, or a mixture of the two plus two bushels of sand and a bushel and a half of clay mixed with water. This is then moulded into balls with the hands and left to dry. Zip firelighters, impregnated with flammable kerosene, were first marketed in 1936. Today's eco-friendly versions are made from a mixture of wood fibre and wax.

OLD VESTS MAKE EXCELLENT DUSTERS

As long as they are made of knitted cotton, which makes them efficiently absorbent. When flannelette sheets were in common use these, too, were regularly pressed into a second life as household dusters.

Before the advent of the vacuum cleaner, and in an era before central heating when homes were lit with open fires, dusting was virtually a daily necessity. Even in the 1930s, this was typical advice: 'A house should be dusted throughout every day. In the living room the fireplaces should be done first and the room swept, and no dusting ought to be attempted until the dust caused by the sweeping has had time to settle. All dusting should be done from the top to the bottom … Every ornament must be moved: it is not enough merely to go round them as that will leave an ugly ridge of dust. Among things which are frequently neglected, yet which provide harbourage of dust, may be mentioned picture frames, bars of chairs and electric light shades or gas lamps…'

Apart from making a room look nicer, regular dusting is beneficial to human health. Even though the dust from coal fires is no longer an everyday hazard to the state of the lungs, dust mites, which feed on the debris shed continuously by the human skin, and the dander from the skins of cats and dogs, are a common cause of allergies.

The saying 'March sun causes dust and the wind blows it about' alludes to typical spring weather and the ability of bright, spring sunshine to highlight accumulations of winter dust in a room.

BETTER DUSTERS

Old tips for improved dusting that are still worth heeding:

A slightly damp duster collects dust more readily.

For a better polish, put a dash of paraffin in the rinsing water when dusters are washed.

A soft piece of chamois leather soaked in cold water and wrung out tightly makes an excellent duster.

Warm dusters are better than cold for achieving a high shine.

SWITCH THAT LIGHT OUT!

Britain's wartime regulation during the blackout, but also a green mantra for today. Modern economy is assisted by the low-energy light bulb, a type of fluorescent light.

To save energy and reduce household bills it makes perfect sense to switch out the lights in rooms that are not in use unless you are coming back into the room within a few minutes. And it is a myth that turning a light on and off reduces the life of the bulb. The low-energy bulb, correctly called the compact fluorescent light, is an evolution of a bulb invented in the late 1890s by Peter Cooper Hewitt for use in photographic work. Its modern form was the brainchild of American engineer

Ed Hammer and it was produced in response to the global oil crisis of 1973. Since then its performance has been much improved to reduce flickering and slow starting.

The stringencies of the blackout began on 1 September 1939, two days before the outbreak of World War II, their purpose being to deprive enemy aircraft of positional information. The law demanded that no chink of light be seen from any home. Car headlights were taboo and even the red glow of a lit cigarette prohibited. Every window quickly became festooned with blackout curtains while ARP (Air Raid Patrol) wardens patrolled the streets ready to report any offenders to the local authorities.

ECONOMICAL HINTS

101 Things to Do in Wartime (1940) included these handy hints for saving on lighting

Light should be concentrated just where it is required most of all.

Lighting should impart an air of cheerfulness.

The light from low-wattage lamps can be considerably increased by fitting reflector shades.

In places such as entrance halls, passages, bedrooms, the lighting may be dimmed, but in the living-rooms, the aim should be good lighting without glare.

A table under a ceiling light would be sufficiently illuminated for handicrafts or table games.

Save buttons from an old shirt

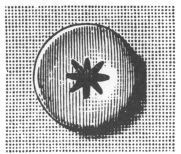

Small white shirt buttons are always useful, and they are well worth keeping if a shirt is worn out and ready to be thrown away. New or reused, buttons properly sewn on by hand should stay in place for as long as the garment lasts.

The correct way to sew on a two- or four-holed shirt button, having marked the exact spot with a pin or tailor's chalk, is to attach the thread to the right side of the fabric with a few small stitches until the thread feels tight (it should not be knotted at the end). Holding the button in place with one hand, insert the needle into one of the holes in the button, then back down again to the wrong side of the fabric. Continue in this way, still keeping hold of the button so that it is not too tight. Once the button is secure, bring the needle to the right side and wind the thread around under the button to make a shank. Then return the needle to the wrong side of the garment and finish with a few small stitches.

To buttonhole a person is to detain them in conversation. The expression comes from the old practice of holding someone by the buttonhole on their lapel while addressing them.

Buttons were invented long before buttonholes. While the garments of the ancients were secured with pins and brooches, buttons were originally items of ornament, made from wood, bone, shells and the like. The oldest ones found to date are about 3,000 years old and come from Mohenjo-Daro in the Indus Valley. The Romans, who did use buttons as fasteners, secured them in place with loops attached to the opposite edge of a garment. After buttonholes were 'invented' in the 14th century, the range of buttons burgeoned, being made from everything from glass to precious gems.

CHIMNEYS SHOULD BE SWEPT ANNUALLY

A wise precaution to keep a chimney clear of all kinds of blockages – from accumulations of soot to birds' nests – and to avert the risk of a dangerous fire, for it is also said that a sooty chimney costs many a beefsteak. The occupation of chimney sweep is associated with many customs and superstitions.

It is inevitable that an open fire will create particles of carbon. Also, if the chimney is poorly ventilated, creosote may collect inside, which, as well as being highly flammable, can cause unpleasant odours. Ideally the chimney should be swept in late summer or autumn before the fire is lit for the first time.

Today's chimney sweep will use brushes very similar to those made iconic by characters such as Bert, played by Dick Van Dyke in the film *Mary Poppins*. However today's home owner will probably be able to manage without the preparations recommended in *The Concise Household Encylopedia*: 'Begin by removing and putting away all white covers in the room. Take everything off the mantelpiece, as the sweep will need it clear to fix his black curtain. Sweep the hearth and grate perfectly free of all ashes and cinders, and spread a thick layer of newspapers all around the fireplace.' In the case of 'an untried man' it also advised taking down the curtains and covering all furniture with dust sheets. And, as a

precaution: 'While sweeping is in progress go outside and see that the brushes go right up the chimney and are visible above it … neglect of this means that soot will fall later.'

The occupation of chimney sweep is an ancient one. For centuries it was customary for sweeps to employ small boys to climb up inside chimneys to clean them with hand brushes. In his *Songs of Innocence* and *Songs of Experience*, published in the 1790s, the poet William Blake powerfully depicted the plight of these 'climbing boys', whose impoverished parents would sell them to master sweeps at around the age of five. Many suffocated in the flues, or died young from cancer and lung disease. Only in 1840 did it become illegal for anyone below the age of 21 to sweep chimneys.

MET FOR LUCK

To avoid ill fortune, take no chances if you encounter a chimney sweep:

If you meet a chimney sweep you should bow to him for good luck, or turn around three times while saying good morning.

The lucky bride is one who meets a chimney sweep on the way to her wedding.

Encountering a chimney sweep on your way to the races will assure you of many winners.

To see the sweep's brush emerging from the top of a chimney will bring you good fortune.

THE RAT SHOULD BE KEPT AS FAR AWAY FROM THE HOME AS POSSIBLE

Indeed it should, not least because rats are the agents of disease as well as being able to damage the home by gnawing through everything from food containers to electric cables. Since milder weather has encouraged their reproduction, nearly all of us now live within a couple of metres of a rat.

'You dirty rat!' is one of the most famous movie misquotes. The nearest James Cagney ever got to it was 'Mmm, that dirty, double-crossin' rat,' in *Blonde Crazy* of 1931.

Even if you do not see a rat you will be aware of its presence. It will leave droppings (larger than those of a mouse) or footprints, or toothmarks on food or food containers, make holes or nests and create rat runs in the garden. As *The Concise Household Encyclopedia* says: 'Very often the intruder has got into the house undiscovered, and announces its presence by unmistakable gnawings and crunchings in the woodwork. It may have visited the larder during the night, and scraps of food dragged about the floor tell their own story.' This cautionary tale illustrates the importance of good housekeeping in preventing rat infestations, but keeping the fabric of the home well maintained is also crucial, as the creatures will make their way in through faults in brickwork.

Rats are hard to eradicate. Traps, poisons and the expert help of a rodent exterminator may be needed. Dogs such as terriers are adept rat catchers, and a cat may also deter rats. According to Buddhist lore a rat once ate part of one of the scriptures of the Enlightened One. So the Buddha took a piece of its skin and turned it into a cat. The proof? That rats are still afraid of cats.

EXTERMINATE ANTS WITH BOILING WATER

A good method for getting rid of garden ants, but not usually practical for ants that have already found their way inside the home. Powdered borax is an effective old-fashioned way of getting rid of them.

When ants have got into the house it may well be possible to trace the route they have taken back to a site in the garden: this is often the lawn, where you may see some tell-tale mounds of fine earth betraying the presence of nests. Pouring boiling water on a mound should kill both adults and unhatched eggs, but it may be helpful to sprinkle some lime over the nest before dousing it.

Indoors, ants may hide in cracks and crevices in the floor and, as *The Household Encyclopedia* correctly observes, take their toll of 'meat, pastry, jam, and dripping. Here,' it says, 'the use of any strong smelling deterrent like carbolic or paraffin is liable to make the food uneatable.' The safe alternative it recommends, which plays to the fact that ants are greatly attracted to sugar, is this: 'A simple syrup sufficiently viscid to clog the ants' limbs will serve as a trap, from which the insects can be plunged into boiling water. The syrup may be made to kill,' it adds, 'by adding a few drops of formalin.'

The industry and ingenuity of ants are acknowledged in the saying from the Bible's Book of Proverbs: 'Go to the ant, thou sluggard; consider her ways and be wise.'

For natural ant control, other substances known to be effective include hot ground pepper, chilli oil, cinnamon, lemon juice and vinegar.

FLIES ARE THE MOST PERNICIOUS INVADERS

And should, of course, be got rid of or, even better, prevented from entering the home, not least because it has been discovered that flies have the potential to transmit more than a hundred diseases to both humans and animals, including typhoid, cholera and dysentery.

> Sailors have long believed that it is a sign of good luck if a fly falls into their drink.

Flies evolved long before humans, but houseflies did not become established as the foes of the householder in Northern Europe until the Iron Age when, around 400 BC, it began to be customary to keep domestic animals indoors during the winter.

Flies transmit illnesses by picking up disease organisms on their leg hairs – from excrement or other contaminated matter – and depositing them on food when they settle on it to eat. The flies' own excreta may also be the source of infection.

Problems with flies are always worst in summer when the insects are most active. And just one female fly can lay more than 500 eggs in her three-week lifespan. These hatch into maggots, which feed on decaying organic matter. Within 10 days each will have pupated and emerged as a new adult, equipped on each of its six feet with hairy pads. These secrete a sticky liquid that enables them to cling to almost any surface.

Before the introduction of fly sprays delivering insecticide in aerosol form, homes were hung about with sticky yellow flypapers. These were most unsightly but economical, especially if made at home. Mrs Kirk provided 1920s housewives with instructions in a flypaper 'recipe' calling for '1oz Resin 1oz Castor oil; a few drops of Honey and a small piece of Beeswax'. The method was as follows: 'Melt together, and spread on firm papers. Try a little first, to see if proper consistency, when cool. If not hard enough add a trifle more resin.' An even older method, said to have been introduced from Italy in the 17th century, was to hang up a cucumber stuck with barleycorns, having first made holes in the cucumber with a bodkin. These were recommended in summer to keep flies from pictures and hangings.

As an alternative, *Cassell's Household Guide* recommended: 'Strong green tea, sweetened well and set in saucers about the places where they are most numerous,' adding the dramatic comment: 'This plan is much to be preferred to the use of those horrible fly papers which catch the poor insects alive, cruelly torturing them while starving them to death.'

BEWARE, BEWARE THE BEASTLY FLY

Some more ways to tackle flies in the home:

Keep food covered or refrigerated.

Hit flies with a fly swatter.

Set out small bowls filled with sweet green tea – flies will be attracted to it and drown.

Spray rubbish bins with vinegar.

Empty and disinfect rubbish bins regularly.

Mix equal quantities of ground black pepper and sugar and add enough cream to make a thick paste. Smear it on to a saucer and put it where flies are likely to appear.

GEESE WILL KEEP THIEVES AT BAY

More trouble than a burglar alarm, but also handy for keeping grass short. Geese can be fattened for the table and their eggs are delicious. On the down side, they are notoriously noisy and messy.

Geese as guardians are nothing new. In the fourth century BC, they were kept to guard the Roman temple of Juno. When, one night, scouts were sent there by Gauls intent on invasion, the geese cackled so loudly that the scouts' cover was blown and the citizens forewarned. In celebration, a goose was sculpted from gold and carried aloft into Rome to celebrate and honour the birds.

'Anyone owning grass land or living near a common,' says *The Household Encyclopedia*, 'will find it profitable to keep geese as, being persistent grazers, they need little in the way of food. Swimming water is desirable when breeding. Their housing,' it explains, ' is quite a simple matter. Any shed or outbuilding sufficing so long as the walls are sound and the floor dry.' A good ratio is four or five females to every gander.

For the table, geese are best fattened during their first year so that they are plump and tender. Although now associated with Christmas, goose was once the traditional fare for the autumn feast of Michaelmas and geese were driven into towns in this season for the goose fairs. In cities such as Nottingham over 20,000 geese would congregate for sale on a single day.

Even a goose of advanced age – they can live to 20 – can produce more than 50 eggs a year. And of course no one should ever think of harming, let alone killing, the legendary goose that lays the golden egg.

The best way to look after books is to read them

Simply because taking a book from the shelf and opening the pages shakes off the dust and keeps the binding flexible and the pages well separated. Books of all kinds need special protection from the ravages of insects.

While they are in the bookcase, books are best kept upright and firmly packed, but not pressed so tightly together that it is a struggle to pull out an individual volume. And because the top of the spine is the most prone to injury, be sure not to tug at and damage it. Precious books need to be kept away from the deteriorating effects of sunlight, which will bleach and weaken cloth bindings, and from extremes of heat and cold. If books are allowed to get damp, mildew and other unsightly and harmful moulds may establish themselves. One old way of discouraging them is to sprinkle a perfumed oil such as lavender oil into a bookcase.

Insects that attack books include the silverfish and the larvae of a variety of beetles. As they feed, these pests turn paper into a kind of lace – if they feed on the thread in the binding a whole book may fall apart. Should larvae hatch inside a volume they will bore their way out, making holes right through the paper. When whole libraries are threatened, fumigation is the last resort, but in the home a natural way to exterminate silverfish is with pyrethrum dust, an organic insecticide widely used in the garden, or with boric acid powder, which is also effective against other book-destroying insects.

On the care of books, *Cassell's Household Guide* is stern indeed. 'Of such acts of vandalism as turning down the corner of a page, or laying

> *It is an old country custom to open a book – usually the Bible – on New Year's Day and to infer from the first words that catch the eye the omens for the year ahead.*

a book open with its face to the table to keep a place, we could wish,' it says, 'it were unnecessary to speak. Of those who persist in doing so,' it continues, 'we can only ask "Is there no such thing as paper, that you may insert a slip?"'

To get a musty smell out of a book, open it and place it in a polythene bag together with 4 tablespoons of bicarbonate of soda (baking soda). Seal the bag and leave it for a couple of weeks.

DRY FLOWERS BY HANGING THEM UPSIDE DOWN

The tried and tested way of producing dried flowers. They also need to be kept in a warm place – damp may cause them to rot. Fresh herbs can be dried in the same way, and make excellent additions to dried flower arrangements as well as being invaluable in the kitchen.

To dry flowers, tie them into small bunches with raffia. They can then be attached to wall hooks or tied on to coat hangers before being hung up. When dried, flowers can either be arranged in a vase or pressed into a block of bone dry florist's foam to make a display of your choice.

A great number of flowers are good for drying but among the best are roses, sunflowers, hydrangeas and gypsophilas. Most grasses also dry well and are very useful for adding 'sculpture' to an arrangement. Other favourites are everlasting flowers or '*immortelles*' such as *Helichrysum*, which, despite their name, still benefit from drying (see box). Many of these are distinguished by having papery bracts. All kinds of dried seed pods and heads look great in dried winter arrangements – pick those of poppy, scabious, teasel and sedum.

Loose flower petals can be dried to make sweet-smelling pot pourri, to which oil-based perfume can be added for extra effect. Leaves, petals and small flowers can also be pressed as keepsakes and used for making decorative cards and collages. If you do not have a custom-made flower press take care when using books for pressing – juices from plant material can cause irreparable damage. Use plenty of thick blotting paper, which will also aid the drying process.

FLOWERS FOR DRYING

All these easy-to-grow plants make superb dried flowers:

HELICHRYSUM – straw flower – pick when outside bracts are open, but not the centre. Stems may need to be wired for use in arrangements.

LIMONIUM – statice – pick as buds begin to open; they will open more as they dry.

SALVIA – blue sage – pick when fully open. Leaves also dry well.

ACHILLEA – yarrow – cut when flower heads are fully developed.

LAVANDULA – lavender – pick when flowers are opening. Also useful for scented lavender bags.

WASTE PAPER HAS MANY USES

Apart, of course, from commercial recycling. Paper is useful for everything from wrapping parcels to lining drawers, so it is always worth having spare paper to hand.

Papier mâché, whose name means 'chewed paper', is made by soaking pieces of paper in paste and sticking them together in layers. If desired they can be laid over a mould or form and can also be reinforced with strips of fabric. Wallpaper paste makes an excellent adhesive. When dry and solid, the material can be painted or decorated as you wish. *Carton-pierre*, a French phrase meaning 'stone cardboard', describes papier-mâché strengthened with chalk and decorated to resemble wood, stone, or metal.

The ingenious Victorians even used 'paper cement' as a material for sculpture, but only for indoor ornaments. *Enquire Within* gives these instructions: 'Take equal parts of paper, paste, and size [glue], sufficient plaster of Paris to make into a good paste, and use as soon as possible after it is mixed. This composition may be used to cast architectural ornaments, busts, statues, &c., being very light, and susceptible of a good polish, but it will not stand weather.'

A nice piece of wrapping paper marred by a grease mark can easily be salvaged by putting blotting paper over it and pressing it with a hot iron. A warm iron applied directly will get rid of any folds and wrinkles.

Lining drawers with paper helps to keep the contents fresh and dry, and tissue paper is good for gift wrapping and for leafing between folds in clothes in drawers and when packing suitcases. Paper stuffed into damp shoes will help to stretch them a little if desired. Spare crêpe or tissue paper can be made into artificial flowers. Petals are shaped and glued together, then fastened with wire, which is then covered with more paper. Equally ingenious is to make paper dolls, complete with clothes, from paper that would otherwise be thrown away.

Good cutlery requires good care

Which means that it needs to be kept clean and well looked after. Silver and stainless steel spoons and forks need quite different

treatment from knives, but all silver needs care to protect it from lingering acids, which can cause it to corrode.

It was only in the 17th century that spoons and forks came into ordinary use in the British dining room. Before that food was speared on the end of a knife or eaten with the fingers. Once forks had been brought from Italy, notably by Thomas Coryat in 1611, they quickly became accepted. It was the development of sheet metal rolling in the 17th century that made mass production of silver cutlery possible – hence its name of flatware. In Sheffield, the home of cutlery making, designs such as rattail were fashioned that remain popular to this day. Following the Restoration in 1660 the fashion for cutlery canteens, already popular in France, crossed to Britain. Good antique sets of silver are enormously valuable: a set of 12 tablespoons and 12 forks from the 1730s in good condition is likely to be worth at least £15,000.

> To prevent silver from tarnishing while stored it was once customary to rub it with olive oil, which would have prevented the air from getting at it.

To keep silver bright and corrosion free, rinse it as soon as possible after use, wash it in up in hot water containing a little washing soda and dry it immediately. If badly tarnished it will need cleaning with silver polish. Choose the thick liquid you apply and buff off with a cloth rather than silver dip and avoid rubbing over hallmarks: if these are smoothed out by excess rubbing, it can diminish the value of good pieces. Stainless steel usually needs no more than washing in hot water with detergent, but for stubborn stains try rubbing on some vinegar. Or you can try a special stainless steel cleaner.

Never soak knife handles in hot water

A mantra for perfect washing up – and essential for any cutlery with handles attached. The simple reason is that hot water loosens the glue used to stick the handle to the knife and discolours, cracks and splits ivory and bone – traditional materials for handles.

Special treatment for knives once extended much further. The Victorian scullery maid would have been under strict instructions to wash blades and handles separately, or to soak blades in a jar of water rather than immersing them, using a little salt to remove stains from knife handles.

Before the advent of stainless steel, which was created in 1913 by the Sheffield metallurgist and cutlery maker Harry Brearley, knife blades were rubbed with mutton fat after they had been washed and dried to prevent them from rusting. Blades were kept bright by rubbing them on a knife board covered in brick dust, which was then carefully removed from the handles with a cloth. By the 1880s rotary knife cleaning machines were becoming commonplace. To quote Mrs Beeton: 'Small and large machines are manufactured, some cleaning only four knives while others clean as many as twelve at once. Nothing, 'she adds, 'can be more simple than the process of machine knife-cleaning.'

The first knives were made from bones, flints and obsidian, a kind of volcanic 'glass'.

Good kitchen knives and carving knives also need to be kept sharp. Knives, with spoons, are the most ancient pieces of cutlery. They were so valuable that until the 16th century a family would share a single knife between them, passing it around the table to cut up food, then eat their meal with a spoon or their fingers, or sup direct from a bowl. Guests would bring their own knives to the table.

Buy perishable goods in small quantities

An essential of careful marketing, even today when perishable foods can be safely stored in the refrigerator. Equally, it is still important to be able to judge foods for their quality and freshness before they are purchased.

Freshness is vital not just for taste but also for health. Green vegetables reduce in vitamin content with keeping, so diminishing their nutritional value, while meat and fish risk becoming infected with agents of gastrointestinal disease. Today, however, matters are made more complicated by the fact that fresh foods such as green vegetables are stored for long periods between harvesting and sale, making it even better to buy vegetables and fruit in season and, if you can, from sources such as farmers' markets.

To save buying more than you need, it is still a good idea to follow granny's example and work out menus for several days ahead, then write a shopping list. And remember that three for two, buy one get one free or similar supermarket 'bargains' will actually cost you money if you end up throwing away surplus ingredients you can't use.

The merits of keeping perishable food cool has long been known. Even before 1000 bc the Chinese were collecting snow

Early home refrigerators, especially in country districts with no mains electricity, were powered by oil or gas. Advice to owners of the 1930s included keeping the appliance spotlessly clean 'or its end is defeated'. It was suggested that it be wiped over every morning with a clean, damp cloth and washed out regularly with a weak solution of soda and water.

in winter and keeping it in cellars for preserving food in hot weather. By the 18th century many large houses in Britain boasted ice houses, but home refrigeration did not become practicable until 1834, when Jacob Perkins patented a 'vapour compression machine', which compressed a volatile fluid, evaporated it to produce a cooling effect, then condensed and recirculated it. He was not, in fact, the first to have this idea. Artificial refrigeration had been demonstrated by William Cullen at the University of Glasgow in 1748, but he failed to use it for any practical purpose.

Buy wine on an apple and sell on cheese

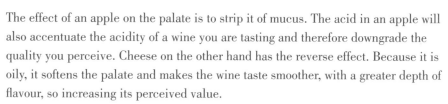

A mantra of household economy that comes from the way that these two foods affect the taste buds and the way a wine feels in the mouth. From a tasting, the value and price of a wine is determined.

The effect of an apple on the palate is to strip it of mucus. The acid in an apple will also accentuate the acidity of a wine you are tasting and therefore downgrade the quality you perceive. Cheese on the other hand has the reverse effect. Because it is oily, it softens the palate and makes the wine taste smoother, with a greater depth of flavour, so increasing its perceived value.

Gone are the days when every house of any size had a wine cellar, but it is still worth taking the trouble to store your wine well. Ideally keep it in a cool dark place where the air is not completely dry and the temperature is as near 10°C (50°F) as possible. Store bottles horizontally – this is especially important for bottles sealed with real corks, as the cork needs to be prevented from drying out. In a house with servants it was the butler's duty to see to every aspect of the selection, storing and serving of wine, as reflected in the proverb: 'The wine is the master's, the goodness the butler's.'

In times past, country people would always make their own wines from home-grown produce such as parsnips, or from blackberries and elderflowers freely available from the hedgerows.

COUNTRY WINES

All kinds of ingredients can be used for home-made wine. The rudiments of the method are exemplified in these two country classics:

BLACKBERRY WINE: Put fruit into a vessel with a tap fitted near the bottom and cover with boiling water. Mash, then leave to stand for 3–4 days until the pulp forms a crust on top. Draw off the fluid, add 500g (1lb) sugar to every 4 litres (1 gallon) of liquid and mix well. Put into bottles, cork tightly and keep for a year to mature. Decant before serving to remove any sediment.

PARSNIP WINE: Wash, peel and finely slice 5kg (10lb) parsnips and cook thoroughly. Strain through a fine sieve. Put the resulting liquid in a pan with 3kg (6lb) preserving sugar and boil for 45 minutes. Add a slice of toast thinly spread with a little fresh yeast. Cover and leave for 10 days, stirring well each day. Strain the juice into a cask (or its modern equivalent) and leave, lightly corked, until it has finished fermenting, then seal it so that it is airtight. Leave for 6–9 months before bottling.

TOO MUCH EMPHASIS CANNOT BE LAID ON THE DANGERS OF DAMP

True, not only because a damp home is unpleasant to live in but because damp can encourage such problems as the growth of unpleasant fungi and more serious problems such as dry and wet rot.

Instructing on the proper construction of a healthy house, *Our Homes* states categorically (and correctly) that it should be built so that 'it should not hold damp in any part. All the materials used in building should be compact, dry, and impermeable to wet. The wood should be sound and well seasoned, the bricks and stones should be free of porosity and power of retaining water; the plaster on the walls, and all the substances with which the walls are covered, should be firm and impermeable.' Added to these precautions, every home needs a proper damp course, which will allow water to drain away, and a good circulation of air to help prevent condensation.

Now that our homes are insulated and double glazed, with windows that are rarely opened in winter, damp caused by condensation is an increasing problem: not only does it cause unsightly black growths of mould, but it can often lead to allergic reactions. It is poorly ventilated bathrooms, shower rooms and storecupboards with exterior walls that are most likely to be affected, because they are most prone to condensation.

Meticulous attention to cleanliness using a bleach-based cleaner is the best prevention, but if fungal growth does appear you should deal with it as soon as you see it. You can buy proprietary treatments, or try removing it with a mixture of 1 teaspoon of water softener, 1 tablespoon of ammonia and 1 tablespoon of vinegar stirred into 200ml (7fl oz) of warm water. After application, rinse the area with clear water then dry it thoroughly.

DEALING WITH DAMP

Some tips old and new:

OLD
Place a box of quicklime in the corner of a cupboard.

Put a block of camphor in each corner of a room, the latter reputedly being 'successful where fires have failed'.

Paint walls with a paint made by mixing 5 parts turpentine, 7½ parts chalk, 5 parts boiled linseed oil and 5 parts resin.

To prevent pictures and mirrors being damaged by damp walls, glue small discs of cork to each corner of the back of the frames.

NEW
Make sure your home is well ventilated – open the windows or fit efficient extractor fans.

Keep lids on saucepans while cooking.

Avoid drying clothes indoors if possible; if you use a tumble dryer make sure it is well ventilated.

Keep your home warm, but don't use paraffin or bottled gas heaters: burning a litre of paraffin puts about the same amount of water vapour into the air.

INDEX